OXFORD MEDICAL

C000301549

Drugs and

Drugs
and
Medicines

A Consumers' Guide

Roderick Cawson MD FRCPath FDS
Emeritus Professor, in the University of London

Roy Spector MD PhD FRCP FRCPath
*Professor of Applied Pharmacology,
United Medical and Dental Schools of Guy's
and St Thomas's Hospitals,
University of London*

OXFORD   NEW YORK   TOKYO
OXFORD UNIVERSITY PRESS

Oxford University Press, Walton Street, Oxford OX2 6DP
Oxford New York Toronto
Delhi Bombay Calcutta Madras Karachi
Petaling Jaya Singapore Hong Kong Tokyo
Nairobi Dar es Salaam Cape Town
Melbourne Auckland
and associated companies in
Berlin Ibadan

Oxford is a trade mark of Oxford University Press

Published in the United States
by Oxford University Press, New York

British Library Cataloguing in Publication Data
Cawson, R. A. (Roderick Anthony)
Drugs and medicines.
I. Title   II. Spector, R. G. (Roy Geoffrey)
615'.1
ISBN 0–19–261655–2

Library of Congress Cataloging in Publication Data
Cawson, R. A.
Drugs and medicines: a consumers' guide/Roderick Cawson, Roy Spector.
p. cm.—(Oxford medical publications)
Bibliography: p. Includes index.
1. Drugs—Popular works. 2. Consumer education. I. Spector, R. G. (Roy
Geoffrey) II. Title. III. Series.
RM301.1.15.C39 1989 615'.1—dc19   89-3137 CIP
ISBN 0–19–261655–2

Printed in Great Britain by
The Guernsey Press Co. Ltd
Guernsey, Channel Islands.

Preface

Many are worried about the harmful effects of medicines. Some also believe that doctors tell them too little or give drugs unnecessarily. Yet other concerns are that medicines may be inadequately tested and that recipients are subjected to needless risks. While there is absolutely no doubt that some concern about the safety of drugs is justified, it is important that this concern should be as well informed as possible and that criticisms of doctors, drugs, or the drug industry should be appropriately directed.

The main difficulty is that the scope of the science of pharmacology and therapeutics—the study of drugs and their actions and uses—is so vast. Some tens of thousands of drugs are available and many of them have a remarkable multiplicity of actions. New drugs are being introduced at a rate which almost overwhelms the medical profession. It would be of little value therefore to list here all these drugs and merely to summarize their uses and toxic effects. Our approach has been rather to outline the principles of treatment of common and important illnesses, to indicate when the use of drugs is appropriate and when it is not, to describe the kinds of side-effect that can appear, and to consider how these side-effects may come about. We begin, however, by looking at the nature of drugs in general and how they may act. In addition we have tried to give some idea of the rigorous nature of testing of modern drugs and give a brief account of the pharmaceutical industry—the scapegoat for so many complaints. Above all, we aim to give the reader a balanced view of the problem of drug safety.

Every effort should be made to ensure the safe use of drugs. Yet doctors, the drug industry, and the drug licensing authorities are not infallible. It is essential in fact to remember that no human acts are risk-free, and if only for this reason, medicines should only be taken when absolutely necessary. Nevertheless, desperate ills call for desperate remedies and sometimes risks must be taken.

The great contribution that modern medicine has made to human health cannot be denied. Indeed medicine may be regarded as being too successful in contributing to the vast overgrowth of the world's population.

Nevertheless, partly because of anxiety about the possible dangers of drugs and partly because of a widespread retreat from science, increasing numbers of people are turning to alternative forms of treatment such as herbalism, nature cures, or health foods.

Despair, anxiety, emotional conflict, and other stresses can certainly lessen the resistance of the body to disease, worsen symptoms, and reduce the effectiveness of drugs. Holistic medicine aims to take into account psychosocial and other 'non-medical' factors underlying illness. Although it is important to look at ways of improving a patient's life situation, it is also essential to recognize the potential value of drug treatment.

To reject modern drug treatment is to return to the Middle Ages when death and disease lurked everywhere and plagues were a constant threat. Then, doctors and their patients had faith in traditional remedies and the sorts of medicine that the witches in Macbeth might have prepared. Faith may play a part in cure but unless a medicine has been extensively and objectively tested over a considerable period there can be little guarantee that it will have the desired effect. Unfortunately, it would also appear that medicine, conventional or unconventional, that has any significant effect on the body, is likely also to have some undesired effects. Therefore all drugs should be used with great care.

The plagues of today are often man-made: epidemics of death and illness are caused by smoking, alcohol, violence, and motor vehicles. The harm caused by modern medicines remains a trifle by comparison.

No reasonable doctor would deny that there have been many failures of modern medicine as well as notable successes. Whatever the defects of conventional medical treatment, much time and quite extraordinarily meticulous care are taken to test the safety of drugs: the same cannot be said for alternative remedies.

Acknowledgements

We are very grateful to Mrs M. Barnett and Miss H. M. Prophet for so assiduously reading the text and for their comments, and to Mr Sidney Luck for his meticulous care in helping us with the library references.

London   R.A.C.
August, 1988   R.G.S.

Contents

1

Drugs: What are they?

The term drug in its strict medical sense includes *any* substance used for the treatment of disease. However, in general usage the word 'drug' is often thought to mean a drug of abuse (addiction). The worry is that a drug will lead to addiction and this fear is greatest in the case of drugs which act on the mind. Yet even among the many drugs which act on the mind, only a few have the propensity for inducing dependence in susceptible persons.

In this book the term 'drug' is used in its broad medical sense, with no implication of sinister properties. The alternative term, medicine, may sound rather more innocuous, but there is no difference between a medicine and a drug.

Some patients resort to herbal products because of their fear of drugs. But any herbs which have an effect on the body do so because they contain drugs. Morphine, for example, is a powerful analgesic (painkiller), derived from the oriental poppy. Moreover, all drugs were originally of natural origin—plant, animal, or mineral—and some important drugs still come from such sources. Quinine (for malaria) and digitalis (for heart failure) are important examples of traditional, plant-derived remedies which have remained in use for hundreds of years. The main difference in the way that they are used now is that the essential active principle is extracted and purified or modified. Precise doses can therefore be given and the effects are more predictable.

However, the great majority of new drugs are synthesized by organic chemists in the pharmaceutical industry. Though, as a consequence, the exact nature of these substances may be known,

1

their effect on the body cannot be so easily predicted. All new preparations are therefore submitted to many tests to determine first, whether they have any effect at all, second, whether they are toxic, and third, the extent of any potential beneficial actions. Frequently, only a single drug emerges after the synthesis of thousands of compounds.

As research continues, increasingly potent and effective drugs are being developed. Most of these substances act by changing in subtle ways the activities of the body's cells or organs and are likely to affect the normal functioning of the body. It is reasonable to assume that the more potent a drug is, the greater its potential for toxic effects. Nevertheless there are wide variations in the risk or severity of side-effects. In the past, if a drug was of any benefit at all, toxic effects were accepted as the price that had to be paid. Newer drugs are increasingly selective in their effect and so less prone to cause toxic side-effects. Nevertheless, apparently valuable drugs are sometimes discarded or their use restricted, because of serious toxic effects that could not have been predicted from anything known of their action in the body and which had not become apparent during testing.

There is an *element* of risk in taking almost any drug. The side-effects may be frequent and severe, or rare and minor but the essential consideration is how greatly the benefits outweigh the hazards.

Where drugs act

One of the more remarkable features of modern drugs is their ability to affect cellular function at astonishingly low concentrations. Many are effective when no more than one part in a million or even one part in a thousand million is present in the blood. This indicates that only a few molecules of such drugs may be needed to influence a cell and suggests that they act on a particular, specialized part of a cell.

Most drugs appear to act on *receptors*, specialized proteins which form part of the surface membrane of cells. These receptors can recognize and respond to drug molecules—the analogy of a lock and key gives some idea of how this may happen. Each cell has many types of receptor and different drugs appear to attach to different receptors.

It seems hard to credit that membrane proteins, which have evolved over the course of millions of years, respond in such a highly specific manner to drugs which may have been synthesized only a few weeks previously. The explanation appears to be that receptors developed alongside the body's natural chemical messengers and that drugs have structures in common with these messenger substances. Examples of chemical messengers are the hormones secreted by endocrine glands, and neurotransmitters released by nerves. Some drugs stimulate receptors in a similar way to the natural messengers; others have an opposing effect and block the action of natural messengers.

An example of a receptor-stimulating drug is salbutamol which is used for the treatment of asthma. Salbutamol resembles the natural hormone, adrenaline which (among other effects) can widen the airways of the lung. It does this by stimulating receptors in the surface membranes of the muscle cells of the fine bronchial tubes. The muscles then relax, causing the airways to open up, lessening the resistance to breathing. This action is valuable in asthma where there are attacks of increased airways resistance causing wheezing or sometimes a terrifying struggle for breath. These attacks can be relieved by salbutamol which acts on the adrenaline-receptors but has a more powerful effect than adrenaline.

Adrenaline also stimulates receptors on other organs. It can act on the heart and cause it to beat faster or even irregularly. This limits its usefulness in the treatment of asthma, and one of the advantages of salbutamol is that it has relatively little effect on the heart. Other kinds of drug can bind to and block the adrenaline-receptors on the heart. This prevents the almost constant stimulation of the heart by adrenaline and similar body chemicals, so that it beats more slowly and less powerfully. One drug which has this blocking effect is propranolol which is used to lessen excessive activity of the heart. Unfortunately, propranolol also binds to and blocks other adrenaline-receptors, particularly those in the lungs. The normal stimuli to keep the airways fully open is thus lost and propranolol causes increased airways resistance. This does not seem to affect normal people but propranolol should not be given to anyone with asthma or bronchitis for whom increased airways resistance is dangerous.

The effects of propranolol on the airways and of adrenaline on the heart illustrate how undesirable (toxic) effects may be inextricably bound up with the normal action of some drugs.

Another characteristic of drugs, related to the precision of receptor binding, is that slight changes in their chemical structure can cause great changes in their effects on the body. Adding an oxygen atom or removing a carbon atom, for example, from a drug molecule, can change or even abolish its effectiveness.

However, not all drugs act on receptors on patients' cells. Antibiotics and related drugs interfere with microbial but not human cells. For example, bacteria have cell walls which protect them from physical damage and the antibiotic penicillin blocks cell-wall formation. Since human cells lack cell walls they are unaffected; penicillin is therefore highly selective in its action and generally a remarkably safe drug.

Toxic effects

Drugs rarely seem to have a single action on the body. Aside from their therapeutic effects (the desired effects), most drugs also have undesired effects, usually termed toxic or side-effects. But even toxic effects, alarming though the term may sound, can sometimes be used to advantage.

The undesirable and sometimes dangerous stimulation of the heart by adrenaline when used to treat asthma has been described. In a different illness, collapse of the circulation due to a severe allergic reaction (anaphylactic shock), this stimulant effect of adrenaline is life-saving. In such emergencies it is therefore imperative to give adrenaline as in such circumstances its benefits far outweigh the risks.

As might be expected, drugs are most toxic in overdose. For example, drugs such as hyoscine (Kwells) or cinnarizine (Marzine), taken for travel sickness, frequently also cause drowsiness. People vary in the level of their response to such agents, but large doses can induce sleep or even coma, and excessively large doses can inhibit areas of the brain which control breathing and the circulation. Respiration or the circulation may be so greatly weakened as to cause the patient's death. In normal doses, travel sickness remedies are harmless enough, but like most other drugs or for that matter, even pure water, can be lethal if taken in sufficient quantities.

Another kind of adverse reaction to drugs is allergy. It is possible to become allergic to many of the complex chemicals in the environment and this includes drugs. Allergy develops because the body tissues when exposed to a substance from outside the body, develop proteins (antibodies; see Ch. 3) which react with that substance. Such reactions cause inflammation which results in itching, rashes or asthma-like difficulty in breathing. Very rarely, allergy can cause death from asphyxia as a result of swelling of the lining of the air passages or from sudden failure of the circulation. Life-threatening reactions of these kinds are called acute anaphylaxis. Those few unfortunate individuals who become seriously ill after a bee-sting are suffering from a form of anaphylaxis. However, many people mistakenly assume that any unpleasant symptoms arising after taking a drug are allergic reactions and this may unnecessarily restrict the use of some drugs in such patients.

A balance always has to be struck between the risks and benefits of drugs, and a doctor would obviously be irresponsible to prescribe a potentially toxic drug for a minor ailment. By contrast, drugs with a risk of serious toxic effects are often given for life-threatening diseases such as cancer or severe infections. Between these two extremes, drugs vary widely in their effectiveness and by no means all of them will cure the diseases for which they are given: often relief of symptoms is the best that can be achieved at present. Moreover, much may depend on the body's ability to overcome the disease and on the patient's will to recover; even so drug treatment is usually essential and may be highly successful.

These then are some of the basic principles underlying the action of drugs, their limitations, and some of the ways in which they can cause undesirable effects. In many cases, the more powerfully a drug acts on the body, the greater its potential for influencing disease and also for interfering with normal functions of the body. Particularly in the past, it could justifiably be said that powerful drugs were powerful poisons. However, as we have shown, drugs can have very specific actions on cells and a major aim of drug research is to use such knowledge to develop drugs which have the least possible unwanted effects. Increasingly this is being achieved, but at the same time it is only too apparent that these complex chemicals can have quite unexpected effects that even extensive and detailed testing fails to reveal.

The conclusions, to be drawn are that modern drugs can be beneficial, to the point in many cases of being life-saving. Often any side-effects are trivial or inapparent but occasionally they are unforeseeable and severe. Drugs should therefore be taken *only* when the benefits are likely to outweigh any ill-effects.

2

Drug development and the drug industry

It is hard to convey an idea of the immensity of the task involved in developing a single new drug and to confirm both its safety and efficacy. Years of work and millions of pounds have usually to be spent. Whatever the limitations of currently available drugs, so many medical requirements have been more or less satisfactorily filled that to invent substantially better drugs has become increasingly difficult. Penicillin, first isolated in 1929, remains a valuable antibiotic. By contrast, interferon, which many thought would be a panacea for viral infections, has taken 30 years of intense effort to make it a prescribable drug. Yet even now its value has not been fully established.

A major problem is the increasing, and often justifiable, anxiety about drug safety. Indeed the demand is sometimes expressed, that drugs should be both *totally* effective and *totally* safe—but ideals of this sort are utterly unrealistic. Quite apart from completely unpredictable actions of drugs, any that have profound effects on bodily processes will (almost by definition) have related actions which may be severe enough to be troublesome.

Despite beliefs to the contrary, every conceivable precaution must be taken to ensure that a drug causes the least possible undesirable effects, but this is an enormous task. Initial tests are carried out on animals. This raises hostility from animals' rights groups because, undeniably, animals are likely to suffer during drug testing. Unfortunately, also, the response of many animals to drugs differ from human responses, so that tests on animals can provide misleading information.

At some stage therefore, the safety and effectiveness of drugs have to be tested on human volunteers. This raises ethical issues and complaints about using people as guinea-pigs. Furthermore, humans also vary from one another in their responses to drugs. Yet another difficulty is that some drugs may cause a significant toxic effect so rarely (perhaps once in a 100 000 administrations) that, to detect them, impossibly large numbers of volunteers would be needed. Unfortunately, therefore, some harmful effects of drugs are not discoverable until they have been in use for a considerable time.

Side-effects of drugs

Side-effects are a major stumbling block in the path of drug innovation. Broadly speaking, side-effects can be divided into those related to the physiological actions of the drug (other than that for which it is normally used) and those which are utterly unpredictable.

Aspirin for example has been in use for over three-quarters of a century and its scale of use (many thousands of tons a year) suggests that it is generally safe. However, it is only relatively recently that its main actions have become understood. Aspirin's action in relieving pain, particularly rheumatic pain, is due to its ability to suppress pain-producing inflammation. Aspirin is the archetypal anti-inflammatory analgesic, and in recent years has spawned innumerable other drugs with essentially similar modes of action.

Aspirin acts by blocking the formation of substances called prostaglandins which are normally produced in the damaged tissues and trigger the process of inflammation. Unfortunately, there are many prostaglandins and prostaglandin products. Some prostaglandins for example protect the stomach lining against its own digestive juices whilst others help to stop accidental bleeding. By suppressing prostaglandin production, aspirin can weaken the stomach's defences against its own acid and may thus contribute to peptic ulceration. Whilst the majority of patients can take aspirin without any ill-effects, in a few it can cause gastric upsets or, rarely, severe gastric haemorrhage. It is therefore unwise to take aspirin if you have previously had a reaction to it or if there is any possibility of having a peptic ulcer.

For most people these risks are small. Moreover, it has been possible to take advantage of the anticlotting action of aspirin in preventing potentially fatal thrombi building up on artificial heart valves. There is also some evidence that taking a small dose of aspirin daily, in the long term, reduces the risk of coronary thrombosis.

Despite decades of research, it is apparent that aspirin also has actions which are not fully understood. After years of controversy it has only recently become reasonably certain that aspirin may play a part in triggering a reaction known as Reye's syndrome, with the possibility of brain or liver damage, in young children. When diseases are as rare as Reye's syndrome it is difficult to exclude mere coincidence and to establish a cause-and-effect relationship. It has to be accepted therefore that an association between a drug and an unusual toxic effect, may not become apparent for a very long time.

Some side-effects of drugs are so bizarre and unexpected as to make it difficult to believe that the drug can be responsible, and sometimes side-effects can even be beneficial.

Benoxaprofen (Opren), an antirheumatic drug (withdrawn because of toxic effects) and minoxidil, used to lower blood pressure, are unrelated in their primary actions and in their chemical structures. However, both have been found to be effective hair restorers.

The first effective drugs for the treatment of depression, the monoamine oxidase inhibitors (tranylcypromine and many others), were discovered when it was noticed that isoniazid, an antituberculous drug, also raised the mood of depressed tuberculous patients. Chemical modification of the isoniazid molecule produced the monamine oxidase inhibitor (MAOI) antidepressants. However, there is another twist to the story. Who could possibly have anticipated that so commonplace a substance as cheese could interact with these drugs (which tend to *lower* blood pressure) to cause a catastrophic *rise* in blood pressure. This could be so severe as to burst blood vessels in the brain. Hardly suprisingly, early reports of this relationship between eating cheese and acute hypertension were simply not believed.

These are but a few of the problems that beset the introduction of new drugs and which have made it necessary to formulate legislation to ensure adequate testing of new drugs before they can be licensed.

Licensing of new drugs

As far as is possible, drugs must meet acceptable standards of *safety*, *quality*, and *efficacy* and these are the terms of reference of the UK Committee on Safety of Medicines. Similar criteria are applied by the Food and Drug Administration (FDA) in the United States.

After initial tests, the essential basis of drug testing is by means of double-blind clinical trials. In these trials, patients are given either the drug under test or a harmless, inactive preparation known as a placebo. Both the drug and the placebo are coded and neither doctors nor patients know who is receiving which preparation. This ensures that no element of suggestion affects the patients' responses.

It is noteworthy that a significant number of people complain of adverse effects from the dummy preparations, rather from the active drug. In addition some conditions get better spontaneously, and it is essential not to ascribe any such improvement to the drug. After completion of the trial, the code is broken and the effects of the drug, compared with the placebo, are assessed. To collect a sufficient number of appropriate patients willing to participate in such trials and who have no excluding conditions (such as pregnancy) is a major exercise. In addition, many laboratory examinations are required to test the effect of the drug on individual organs.

Such trials, by virtue of their double-blind nature are as objective as it is possible to achieve. They are the acid test which distinguishes orthodox medical drug treatment from the many forms of alternative medicine. There is no doubt that some herbal remedies (for example) may be effective in appropriate conditions, but *objective* evidence as obtained from double-blind trials is lacking.

A licensing body such as the UK Committee on Safety of Medicines to insist on and to assess the results of such trials, is essential and the work that these bodies do is invaluable, but they cannot be expected to be infallible—some of the reasons for this have been touched on earlier. Moreover, there is a heavy price to pay in terms of meeting the costs of testing during the development of a new drug. The documentation, alone, supporting a drug license application can comprise so many volumes as (without exaggeration) to fill a moderate-sized room. It hardly seems possible to check so much data in all its details. Nevertheless, no drug can be prescribed until it has been tested and licensed in this way.

In addition to the Committee on Safety of Medicines and its subcommittees, doctors, the Pharmaceutical Society and pharmacists are on the look out for toxic effects of drugs. There are also regular publications, particularly the *Drug and Therapeutics Bulletin* (published by the Consumers' Association) and the associated *Adverse Drug Reaction Bulletin* which are sent out to all doctors. Doctors and dentists are asked to report to the Committee on Safety of Medicines anything that they suspect might be an adverse drug reaction.

Despite these monumental efforts, it occasionally happens that a drug which has met all the established criteria proves, in practice, to be unsafe for some people. Apart from any other considerations already discussed, there are wide variations between individuals in their response to drugs. These variations depend on genetic make-up, age, racial background and many other factors. As a result it is not possible to have a truly representative sample of any population on which to test a drug.

The elderly, for example, tend to be in greater need of drugs than the young, but they are also more prone to drug reactions, often as a result of deteriorating kidney or liver function. Since drugs can only be satisfactorily tested on completely healthy volunteers, elderly persons are almost by definition excluded from such tests.

Another problem is that the testing of a drug and the assessment of these tests by the licensing authority often result in considerable delay before the drug can come into use. The consequences can be unfortunate: hundreds if not thousands of lives were probably lost prematurely in the United States as a result of the much slower approval by the FDA, of beta-blocking drugs (Ch. 5) which reduce the mortality from heart attacks.

The drug industry

Opinions and emotional responses towards the drug industry can be so extreme that it is sometimes impossible to hold a calm discussion about the subject. Some of the attacks on the industry are based on such catch-phrases as 'health must not be used to make profits'. Since, world-wide, malnutrition is the chief cause of disease, the same complaint could be made against the food industry. As to the matter of profits, it is difficult to imagine how otherwise, the billions of pounds necessary for drug research could be found. Some pharmaceutical companies—the Wellcome Foundation is a major example— plough back all their nominal profits into research. A significant proportion of Wellcome's profits are used by the Wellcome Trust, a charity that is totally independent of the parent drug company and finances research unrelated to any Wellcome product. Nevertheless, the cost of drug development has become so high that even the Wellcome Foundation has recently been forced to raise additional capital on the stock market.

Those, who reject conventional drugs as being foisted on them by a profiteering pharmaceutical industry, seem blissfully unaware of the size and profitability of the 'health food' industry, of the specious claims made by some of its members, and of the emotive advertising that is used. In the United States alone a gullible public spends between four and five *billion* dollars a year on unproven and ineffective cancer 'cures' and about twenty-five *billion* dollars a year on miracle remedies that range from simple lemon juice to milk from 'specially treated' cows.

This is not to minimize the fact that some pharmaceutical companies do make substantial profits for their shareholders, but even the pharmaceutical giants do not succeed in this every year and those companies which make losses are conveniently forgotten. Investors have to be encouraged to finance research and development, and it is, incidentally, very difficult to think of any useful drugs that have come from any other countries than those which depend on this much vilified capitalist system.

The pharmaceutical industry (profitable or not) really does make a contribution to human health. By comparison, the motor industry, far from promoting health, is indirectly responsible for over 5000 deaths and 100 000 permanently disabling injuries each year in Britain alone. The British motor industry has been massively subsidized by the taxpayer and there is hardly a whisper of criticism of the vigorous advertising of cars some of which are capable of exceeding twice the legal speed limit and which encourage the aggressive impulses of adolescent-minded tearaways.

How often do drugs cause toxic effects?

A common criticism of the drug industry is that it (or doctors) fail to tell patients enough about the side-effects of drugs and how often they happen. In fact, drug companies produce *Drug Data Sheets*, which provide, in exhaustive detail, information about their products and all known or even suspected side-effects. The difficulty is that it is hardly ever possible to know how often any of these drug reactions are likely to be seen.

The frequency of drug reactions can only be assessed in relation to the frequency of drug use. Unfortunately this information is virtually impossible to obtain. The UK Department of Health can (with difficulty) discover the number of prescriptions that has been handed out for any given drug, but this does not give an accurate indication of drug consumption— inspection of many bathroom cupboards will show more than a few bottles of unconsumed tablets. That aside, it is often difficult to distinguish effects which are genuinely caused by a drug from purely coincidental illness.

Even if it were possible to tell a patient the statistical chances of suffering any one or more of 60 possible toxic effects of a drug, it would hardly be helpful. For example, it is small consolation to someone who has been struck by lightning to be told that there was little chance of it happening.

Finally, the possibility cannot be ignored that pharmaceutical company representatives may, not necessarily consciously or deliberately, mislead doctors about the effectiveness and safety of products about which they inevitably feel enthusiastic. This may not be done deliberately but there have been occasional black sheep among the many pharmaceutical companies world-wide. Perhaps this is a price that has to be paid for innovative industry, whether it makes drugs or anything else.

Conclusions

First, the development of new drugs depends not merely on gifted research workers, but also on entrepreneurs prepared to take a chance with the vast capital needed for development work and testing. This is not something that a University research department can do. For example, although the initial discovery of penicillin was made in a medical laboratory, its development in quantity as a useful drug depended on persuading drug companies to take on the task and the stimulus of a world war.

Second, drugs, by their very nature, cannot be 100% safe. Do not be misled into thinking that natural substances (herbal remedies) are any better or safer. They may do what is required, but unlike prescribable drugs, they have not been tested for either safety or efficacy. Moreover, some of the most powerful poisons are natural products.

Third, the toxic effects of some drugs may be severe but so infrequent as not to become apparent for a very long period— sometimes for decades. There is no way of testing for such rare reactions.

Fourth, standards for the safety, quality, and efficacy of drugs must be established and this can only be done by an expert official body, independent of the drug industry. In the UK, this body is the Committee on Safety of Medicines. Nevertheless, even the most expert and impartial committee cannot be expected to be infallible. Moreover, such an organization with its advisors, medical and legal, and its supporting staff, inevitably forms an expensive bureaucracy.

Fifth, the cost of drugs must take into account the expense of the development work and testing. An expensive drug may more than justify its cost by eliminating the need for a patient to be admitted to hospital. Admission to hospital is unlikely to be enjoyable for the patient and will probably mean loss of earnings; it also costs the taxpayer hundreds of pounds a day for each admission.

If there is any doubt about such a statement, it is only necessary to be reminded of the introduction of the National Health Service, in 1948. It was planned at that time to spend millions of pounds on additional tuberculosis sanatoria. By coincidence, in that same year, the antituberculous drug streptomycin became available. This saved the expense of building the sanatoria, and also staffing and running costs. The antituberculous drugs alone have therefore not only preserved many lives but also saved the taxpayer millions of pounds.

Sixth, the pharmaceutical industry, aware of the vulnerability of its reputation in the public eye, generally does its best to keep up standards among its members.

Finally, even though unwanted effects of drugs are more common than we would wish, they are rarely as frequent as some seem to believe. If we are ill we have to take some chances in order to get better—as we do in virtually any activity. It is essential to acquire a sense of proportion about the risks involved. We recall a patient who was frightened of taking a virtually harmless drug but who was perfectly happy to ride a motor cycle capable of speeds up to 150 mph: he eventually appreciated the paradox.

The lesson is that, despite all efforts, no drug or anything else produced by human hands is perfect. Neither the most expert medical supervisory bodies nor the pharmaceutical industry can be infallible. Medicines should therefore be taken *only* when they are really needed and in the knowledge that they *may* have some unwanted effects. In this imperfect world we cannot have it all ways.

It is particularly unjust, as so often happens, to remember only the toxic effects of a particular drug and forget the thousands of patients it has benefited. The media are largely interested in bad news and the creation of sensations—real or imaginary. Capital can readily therefore be made from stories (sometimes dramatized or exaggerated) about the toxic effects of a drug, but they frequently ignore such aspects as the rarity or unpredictable nature of such events, or the value of the drug in other respects.

3

Infections, antibiotics, and immunization

The remarkable effectiveness of penicillin and other antibiotics tends to overshadow the many infections, particularly those caused by viruses, for which there is no specific, effective treatment. An enormous variety of microbes is capable of causing human disease. These include viruses (invisible under a conventional microscope), bacteria, protozoa (such as the organism causing malaria), and parasitic worms.

Nowadays, antibiotics can quickly overcome many bacterial infections. However in the nineteenth century major epidemics of cholera (a source of terror in Victorian London) and other water-borne infections which killed hundreds of thousands of people each year in Britain were eliminated by the introduction of effective sanitation before it was known that these diseases were caused by bacteria. Although the notion of keeping excreta out of the water supply largely eliminated these infections, the technical difficulties of doing so in large cities were formidable. Tuberculosis, which is mainly spread from person to person, was also declining before the discovery of antibiotics as a result of better living standards and reduced opportunities for the bacteria to spread. The impact of the antituberculous antibiotic, streptomycin, in about 1948, and its successors was to reduce the mortality from tuberculosis among those who had already acquired the infection. Infections may therefore be preventable without the intervention of drugs, and this is the chief hope for many of the diseases caused by viruses for which no antibiotics or other drugs are effective.

17

The nature of antibiotics

Antibiotics (strictly speaking) are chemical substances produced by bacteria or fungi, but which kill or prevent the proliferation of other microbes. They are highly selective in this activity, by blocking specific aspects of a microbe's life processes. Penicillins and cephalosporins, for example, block the formation of bacterial cell walls, without which the bacteria cannot live. Many other antibacterials interfere with bacterial DNA formation. Since, antibiotics affect specific microbial functions they should be non-toxic to humans. Any toxicity is incidental and frequently unrelated to antibacterial activity. The highly specific modes of action also mean that antibiotics are active at remarkably low concentrations, sometimes at a dilution of one in a billion in the tissues.

A distinction is sometimes made between *antibiotics* and similar, but entirely synthetic drugs (such as the sulphonamides) called chemotherapeutic agents or *antibacterials*. However, many antibiotics are greatly modified for medical use (there are now countless penicillin derivatives with different ranges of activity), whilst entirely synthetic agents can be just as specific in their modes of action as antibiotics. Penicillin was the first effective antibiotic in the strict sense of the word but the first selective antibacterial drug was sulphanilamide, one of many sulphonamides which ushered in the antibiotic era.

Sulphanilamide is a relatively simple chemical which interferes with bacterial cell processes in a specific manner. Graphic descriptions have been given of its effect on pneumococcal (lobar) pneumonia in the 1930s. At that time lobar pneumonia was a common infection which attacked and frequently killed many previously healthy young men. Almost overnight the picture changed. Patients for whom there had previously been little hope, when given sulphanilamide, made what seemed then to be a miraculous recovery. The conquest of streptococcal and several other previously lethal infections with sulphonamides quickly followed.

The sulphonamides still have their uses, but their value has greatly declined owing to the development of resistance to them by bacteria. Overuse of antibiotics has led to the proliferation of resistant bacteria, especially in hospitals, and some bacteria no longer respond to available antibiotics. Although bacterial infections are much less of a threat now, there is a need to develop new antibiotics to deal with resistant bacteria.

Antibiotics are generally safe drugs, but they are by no means harmless. Penicillin seems particularly innocuous, especially when one considers that it has been in continuous use for nearly half a century. Nevertheless people allergic to penicillin can develop rashes or other reactions. On rare occasions it can provoke acute anaphylactic shock which, if not treated immediately, can be lethal in less than five minutes. This is an extreme case, but no antibiotic (or any other drug) can be guaranteed free from adverse effects. These adverse effects and the risks of promoting bacterial resistance are a compelling argument against the indiscriminate use of antimicrobials (particularly for the common cold and other viral infections) just in case they may do some good.

Incidentally, *antibiotics* and *antibacterials* are quite different from *antiseptics*. Antiseptics are little more than poisons—phenol (carbolic acid) is a classical example—which happen to be toxic to micro-organisms but which are just as likely to kill human tissues. They are of no use for systemic (internal) infections as they are too poisonous, only effective in high concentrations, and act very slowly. Antiseptics can be useful for superficial infections, such as ringworm (a fungal infection), where they are directly applied to the infected area and the skin prevents significant absorption. Antiseptics are also used for cleaning floors and work surfaces in hospitals and particularly operating theatres to prevent the spread of bacteria and contamination of wounds. Disinfectant is an alternative term sometimes given to antiseptics when they are applied to inanimate objects, but they are the same substances. However, antiseptics are not effective for sterilizing surgical instruments: this must be done by heat (usually autoclaving in high-pressure steam). The habit of swallowing antiseptic permanganate of potash solutions to ward off gut infections in the tropics, did no real good.

The use and abuse of antibiotics

It is likely that La Traviata and La Bohème would not have died so lingeringly and 'romantically' if they had received competent modern medical management and the appropriate combinations of antibiotics.

Some examples of infections for which antibiotics must be given and may be life-saving are severe streptococcal and staphylo-coccal infections (such as septicaemia), meningococcal meningitis, infective endocarditis (a heart infection), pneumococcal pneumonia, tuberculosis, acute osteomyelitis (a suppurative bone infection), infected burns, and enteric (typhoid) fever. Many of these are discussed in later chapters.

For antibiotics to be fully effective, it may be necessary to identify the organism causing the trouble; otherwise an inappropriate choice of antibiotic may be made and be of no benefit. In addition to the help given by antibiotics, normal immune defences are usually necessary to overcome infections. People who are seriously immunodeficient, i.e. lacking in normal immune defences, (such as patients with AIDS) usually die from overwhelming infections which are unresponsive to treatment.

Some examples of common infections may help to clarify the reasons when and why antibiotics should be used or avoided.

Sore throat

Sore throat is most often caused by viruses for which there is no effective treatment. Antibiotics are more likely to do harm (such as allergic reactions) than good under such circumstances and should be avoided.

Occasionally, sore throat is caused by streptococci. This is usually a mild infection and though highly responsive to penicillin there is rarely any justification for giving it. In a few people, infection can lead to rheumatic fever. This only becomes apparent after the first attack and long-term treatment with penicillin is needed to prevent further streptococcal throat infections. Partly as a consequence of this measure, (but several other factors are involved) rheumatic fever has become a rare disease in developed countries. Severe streptococcal throat infections leading to abscess formation round a tonsil (quinsy) are also rare but need immediate antibiotic treatment.

Diphtheria

Diphtheria used to be a common and important cause of sore throat. The diphtheria bacilli multiply in the throat, causing ulceration. The disease results not from bacterial invasion itself, but from the formation and release by these bacteria of a powerful toxin which can paralyse nerves and stop the heart. Though diphtheria bacilli are sensitive to antibiotics such as erythromycin, this will not counteract the effect of the toxin. Treatment with antitoxin is more important, although an antibiotic may help to prevent further proliferation of the bacteria.

Diphtheria is preventable by immunization and has largely disappeared.

Agranulocytosis

This rare disease is an important example of sore throat (among other symptoms) being *caused* by certain drugs. These drugs damage the bone marrow so that it cannot produce the white blood cells (granulocytes) which help to defend the body against infection. This allows bacteria to proliferate and cause severe ulceration in the mouth and throat. Antibiotics which can cause this complication are sulphonamides and chloramphenicol. Since marrow poisoning by chloramphenicol can be irreversible and fatal, this drug is now a reserve antibiotic and given only for life-threatening infections which are unresponsive to other antibiotics.

Even though agranulocytosis may occasionally be caused by antibacterial drugs, the infections which subsequently develop need to be controlled by other antibiotics such as penicillin which is not harmful to the bone marrow.

Food poisoning

Food poisoning in the general medical sense is a gut infection caused by bacteria or their toxins present in food or water. Food poisoning is particularly common where the standards of hygiene are poor and in hot countries. Although food poisoning is an infection, antibiotics are not, in general, an appropriate form of treatment. They are often ineffective, and many of the antibiotics that act against such infections can themselves cause gut disturbance and diarrhoea. Replacement of lost fluid is far more important in the immediate treatment of diarrhoea.

An extreme form of this type of disease is the water-borne infection, cholera. Despite the terror which cholera justifiably provokes, its effects can be adequately controlled by fluid replacement in the form of a dilute solution of essential salts and sugar. An antibacterial, such as tetracycline, may shorten the infection, but it will not prevent death from dehydration. Before this was understood countless millions have died from the infection. Cholera still causes many deaths in the tropics as a result of delayed or inappropriate treatment and frequently also malnutrition.

A digression on immunity and immunodeficiencies

The spread of AIDS (acquired immune deficiency syndrome) has brought home the importance of the immune system in protecting the body. Although antibiotics are invaluable drugs, normal immune defences are usually also necessary to overcome infections. Severe immunodeficiencies such as AIDS are usually lethal because the person affected is susceptible to overwhelming infection by a great variety of microbes, and this can no longer be controlled by antibiotics.

Immunosuppressive treatment, essential for most patients who require kidney or other transplants to enable them to tolerate the new organ, has somewhat similar but generally less severe effects. This treatment stops the body from recognizing the new organ as a foreign body and rejecting it, but at the same time lowers resistance to other foreign bodies, such as microbes.

It seems appropriate to say something about the complex subject of normal and abnormal immune responses.

Immunity

Everyone is constantly exposed to microbes capable of causing disease. Frequently, these microbes fail to gain a foothold or recovery takes place without the help of drugs. If this were not the case, life would have been wiped out long before antibiotics were introduced.

Natural defence mechanisms have evolved in all animals. These are both non-specific and specific mechanisms of defence. Non-specific defences include barriers such as intact skin and normal stomach acid, which kills many swallowed microbes. Some microbes can get past these barriers or may enter the body via wounds. Once within the body, however, they meet the specific immune defence system. This system involves special cells in blood and in tissues which can recognize substances that are in the slightest way foreign to the body. These foreign substances (such as those in or on microbes) which alert the immune system to respond are known as *antigens*. The immune system responds to antigens by producing *antibodies* which circulate in the blood. In addition a variety of white cells are activated to attack and if possible destroy and eliminate the foreign cells or other material.

White cells are, incidentally, so called to distinguish them from the red blood cells which are coloured by the oxygen-carrying pigment, haemoglobin. There are several types of white cell, but lymphocytes are absolutely crucial participants in the immune defences. Neutrophil granulocytes are less specialized and, so to speak, the infantry of the body's defences.

The immune response is immensely complicated and if it goes wrong can cause harm in various ways. Disorders of the immune response fall into two broad categories: first, responses can be inadequate and give rise to the immunodeficiency diseases, among which AIDS is an extreme example; second, immune responses can be disordered or excessive. The results are the immunologically mediated diseases, such as rheumatoid arthritis and the various allergies, such as asthma.

Drugs and immunity

Drugs can affect the immune response in various ways. They can, for example, trigger allergic reactions. Allergy to penicillin is one of the most common and often the most severe reaction but there are many others. Other drugs depress immune responses. The chief examples are the immunosuppressive drugs which allow the body to tolerate transplanted organs. Corticosteroids used in the treatment of many diseases are also immunosuppressive and increased susceptibility to infection is a serious side-effect of their use. Most anticancer drugs are also immunosuppressive in that they destroy cells of the immune system.

Vaccines are a quite different type of drug which activate normal immune responses, while antisera supplement them by means of sera containing preformed antibodies. Unfortunately there are no drugs which fully restore defective immune responses. AIDS for example is untreatable in this way.

Immunization: vaccines and their uses

In the eighteenth century, Jenner noticed that milkmaids who had acquired cowpox, a mild infection, from cattle did not appear to suffer from smallpox, at that time an epidemic disease with a high mortality. By inoculating fluid from the milkmaids' skin blisters (pocks) Jenner was able to confer immunity against smallpox in others.

There are hardly any other infections where so successful a preventive measure has been so readily found. Most vaccines have been difficult to prepare and prone to cause side-effects which have limited their use. Such are the difficulties that safe and effective vaccines have been produced for relatively few diseases.

Vaccines contain the specific antigens of a particular microbe that stimulate the immune defences into producing antibodies or activating immune system cells. They 'deceive' the body's defences into responding as if the actual infection had been acquired. Vaccines are derived from a harmless variant of a microbe, killed microbes, or an extract of the microbe. The problem is to make a preparation which is harmless but nevertheless stimulates the appropriate immune response. In some infections, notably influenza, the virus changes its character so frequently that preparation of vaccines cannot keep pace.

The medical term for vaccination is *active immunization* since the body's own defenses are stimulated into action. Its effects are long lasting but because this kind of protection takes some weeks to develop vaccines need to be given before an infection is acquired. *Passive immunization* is achieved using antisera, which already contain antibodies obtained from animals or people who have been exposed to the infection. Antiserum gives immediate protection and can be life-saving in some diseases, such as diphtheria and tetanus.

Though prevention is better than cure, it is only justifiable to use vaccines if the disease is a serious threat to health. For example, the need for vaccination against mumps has been questioned since the infection rarely has severe or permanent consequences. But for some infections no reliably effective treatment is available. Tetanus (lockjaw) is treatable if caught early enough and if suitable facilities are available; nevertheless it has a mortality of 30 per cent. Immunization with tetanus toxoid is sufficiently safe to be given routinely to children as the standard DPT (diphtheria, pertussis, and tetanus) triple vaccine, and only requires booster doses at long intervals to confer excellent protection. Rabies is an example of a lethal and untreatable disease where immunization is essential when there is risk of exposure to it.

In the case of less dangerous diseases, the decision as to whether or not immunization is necessary can be very much more difficult. Much publicity has, for example, resulted from cases of brain damage (encephalitis) which may have been caused by vaccination against whooping cough (pertussis). Although it has been established that the risk of brain damage from the vaccine is low in comparison with that from the disease, the acceptance rate for whooping cough vaccination has dropped alarmingly. The result has been a sharp increase in the incidence of whooping cough and its complications. Further, the public has become very wary of vaccines in general. There is now a reluctance to accept measles vaccination for children. Measles vaccination is, however, essential as the disease can have serious consequences such as encephalitis or bronchopneumonia and, world-wide, has a significant mortality. Extensive use of the vaccine has virtually eliminated endemic measles in the USA and there has been a negligible incidence of complications. The capricious nature of this attitude to vaccines is shown by the urgent call for a vaccine against AIDS. If such a vaccine were produced, would there then be complaints about any complications it might cause, conveniently forgetting the protection it might give?

Immunodeficiency diseases

The essential feature of immunodeficiency diseases is an increased susceptibility to infections as a result of failure of the immune system to work as it should. Frequently the microbes causing these infections are harmless to normal people—and known as opportunistic organisms.

Occasionally immunodeficiency is an hereditary disorder and can be so severe that death may come in early childhood as a result of uncontrollable infections. Often the only hope for children with hereditary defects is to have a transplant of normal bone marrow so that immunologically active cells can be formed.

Most immunodeficiencies, however, are acquired either as a result of natural disease or of treatment with immunosuppressive drugs. World-wide the most common causes of immunodeficiency are malnutrition and malaria. Many viral infections cause some degree of immunodeficiency. Important examples are measles and influenza which leave the patient so susceptible to other infections that, in the past particularly, bronchopneumonia was a commonly fatal complication.

The acquired immune deficiency syndrome (AIDS)

AIDS is the most severe, naturally acquired immunodeficiency. It results from infection by the human immunodeficiency virus (HIV) which directly attacks the lymphocytes (a type of white blood cell) that play an essential role in fighting infections. The result is susceptibility to a bewildering array of infections by bacteria, viruses, and fungi, many of which are harmless to uninfected people. Many of these infections do not respond to the antibiotics now available. Although AIDS patients are also unusually susceptible to various tumours, infection is the most common cause of death. Treatment with an antiviral drug, zidovudine (Retrovir®) blocks the proliferation of the virus and in the short-term produces clinical improvement. However, zidovudine has a variety of toxic effects and its long-term value is questionable—it is certainly not regarded as a cure. There is hope that a vaccine may be produced but this will be a difficult task. Prevention by reducing spread of the disease is therefore of the utmost importance. The disease is mainly spread by promiscuous male homosexual activity, sexual contact with bisexual males, and intravenous drug abuse; other means of transmission exist but are far less important.

Immunosuppressive treatment

Immunosuppressive drugs are used for the treatment of severe auto-immune diseases such as lupus erythematosus and for allergic diseases such as asthma, unresponsive to other forms of treatment. In addition, immunosuppressive drugs are usually essential for suppressing the normal immune responses which cause organ grafts to be rejected. The most commonly used immunosuppressive drugs are the corticosteroids. Cortisone is a well-known example but another, prednisolone, is more effective for most purposes and more widely used.

The mode of action of these drugs on immunity and inflammation is complex and by no means fully understood. They probably act by suppressing lymphocytes and also the synthesis or release of a variety of chemical substances (mediators) that initiate inflammation and some immunological activities. Because of such activity the chief complication of corticosteroid treatment is abnormal susceptibility to infection but there are a great variety of other adverse effects, as discussed in Chapter 10. There are many anti-allergic and anti-inflammatory corticosteroids which differ chemically and in their actions to a greater or lesser extent.

In addition to this deliberate suppression of immune defences, some drugs can occasionally have a similar effect by poisoning the bone marrow where the defensive cells are produced. Examples are chloramphenicol (a rarely used antibiotic), sulphonamides (occasionally), phenylbutazone (an antirheumatic drug) and antithyroid drugs such as carbimazole.

Immunostimulants

As yet, there are no drugs which completely fulfil this desirable role. The most promising candidates are the interferons whose main role is in combating viral infections, but they also affect immune defenses to some degree. These drugs have only very recently become generally available and the evidence of their value in either role is limited. Moreover, their toxic effects— particularly a persistent feverish, flu-like illness—are severe.

A review of the methods of controlling infections

Infections are a common a cause of illnesses which even now can be fatal, and sometimes spread widely with catastrophic consequences. Many methods have had to be devised to deal with such problems.

Prevention

Avoidance of exposure to as many microbes as possible is the ideal approach. This is often impractical in the case of such nuisances as the common cold, a continual hazard for commuters, for example. By contrast, other infections, such as the sexually transmitted diseases can normally be avoided by a little restraint in following wayward impulses.

Vaccination is particularly valuable for serious childhood infections and for infections where there is no reliably effective treatment. For some dangerous infections, such as meningococcal meningitis or infective endocarditis, it may be feasible to give a course of an appropriate antibiotic when patients are likely to be exposed to the microbes that cause these diseases. Antibiotic prophylaxis, however, has relatively few applications.

Antiseptics have very limited value in the prevention of infection. They are much too toxic to be taken internally and they are not effective for sterilizing surgical instruments. At home there is little benefit (except to the manufacturers of course) from pouring antiseptics down the drain or into lavatory pans—this usage probably stems from the Victorian belief that diseases were spread by nasty smells.

Treatment

Antibiotics are effective for the treatment of many infections, but by no means all. They are useless for common viral infections such as colds, they can sometimes do more harm than good, and they are in no sense tonics. There are as yet few safe and effective antiviral drugs, but acyclovir for herpetic infections is an outstanding example.

When for example an abscess forms as a result of infection, surgical drainage of the pus is sometimes as important or more important than antibiotic treatment. Toothache or dental abscesses are common examples of infections where antibiotics are rarely of any great value. To extract the tooth or drain the infection by other means is more effective. Nevertheless, patients frequently suffer needless days of pain or facial swelling following antibiotic treatment which acts only slowly or is inactive against the causative bacteria.

In the treatment of toxic infections, such as diphtheria or tetanus, antitoxins are far more effective than antibiotics.

Some important antimicrobial drugs and their uses

This is a vast subject not only because of the number of these drugs but also because they have overlapping activities. When more than one antibiotic may be effective against a particular infection a choice has to be made by considering which of them is least likely to have toxic effects and which is least unpleasant for the patient in terms of ease of administration.

Penicillins

Benzylpenicillin (the first penicillin; penicillin G) has to be given by injection but is still immensely valuable for severe infections by most streptococci, pneumococci (lobar pneumonia), meningo-cocci (meningitis), some staphylococci (but many are now resistant), and the sexually transmitted diseases syphilis and gonorrhoea. Increasing numbers of gonorrhoeal infections are now, however, resistant to this penicillin and have to be treated with spectinomycin. It is also ineffective against the Gram-negative bacilli which mainly infect the gut and urinary tract.

The chief *risk* from benzylpenicillin and all other penicillins is that of allergy. The main *limitation* of benzylpenicillin is that many bacteria have developed resistance to it. It is largely inactivated by gastric acid and absorption from the gut is low; therefore it is best given by injection. Injected penicillin is also very rapidly excreted in the urine and therefore has to be given at frequent intervals. Otherwise, this penicillin is a remarkably safe drug apart from the risks of allergy.

Phenoxymethyl penicillin (penicillin V) is not destroyed by gastric acid and can be given by mouth. It has similar uses and limitations to benzylpenicillin but is less effective for severe infections as it may not be possible to achieve sufficiently high blood concentrations.

Broad-spectrum penicillins differ from the earlier penicillins in that they are effective against many Gram-negative bacillary infections but are not useful for typhoid fever or food poisoning. Their main uses are for exacerbations of chronic bronchitis and middle ear infections (which are usually due to many different microbes) and urinary infections. These penicillins are given by mouth. Ampicillin was the earliest example but amoxycillin is better absorbed, even when there is food in the stomach; both can be given by injection if the need arises. In addition to the risk of penicillin allergy, this group of penicillins is prone to cause irritating rashes unrelated to allergy. They are also destroyed by the enzyme penicillinase produced by staphylococci and various other bacteria, but resistance to this enzyme can be conferred by the addition of clavulanic acid. The latter in combination with amoxycillin is available as Augmentin.

There are also penicillinase-resistant penicillins. For example, cloxacillin and flucloxacillin are resistant to staphylococcal penicillinase. Staphylococci are common and dangerous bacteria which cause a great variety of infections and are very readily spread, especially in hospitals. They typically cause suppurative (pus-producing) infections such as boils and carbuncles, bone and joint infections, septicaemias (bloodstream infections), surgical wound infections, and bronchopneumonia, particularly in the elderly. Some staphylococci in hospitals have developed resistance to all available antibiotics—this is mainly the result of overuse of antibiotics which encourages the bacteria to adapt to them and then, by destroying competing bacteria, opens the way to the establishment of resistant infections. The penicillinase-resistant penicillins are valuable virtually only for staphylococcal infections.

In addition to the examples already mentioned there are now many special penicillins for use against particularly troublesome bacteria, such as pseudomonads which infect burned and immunodeficient patients, and are difficult to eradicate.

Cephalosporins

These antibiotics are chemically very similar to the penicillins and hence about 10 per cent of those allergic to penicillin are also allergic to the cephalosporins. Cephalosporins have many theoretical advantages over the penicillins in that they are more resistant to penicillinase and have activity against a broad range of bacteria. In practice, however, there are few indications for their use other than for infections unresponsive to other antibiotics.

As with the penicillins, allergy is the chief hazard but one of the cephalosporins, latamoxef, is prone to cause abnormal bleeding. Examples of the cephalosporins include cephradine, cephazolin, cefuroxime, and cephamandole.

Tetracyclines

The tetracyclines have the broadest spectrum of antibacterial activity of any of the antibiotics but increasing bacterial resistance has greatly lessened their usefulness. Unlike the antibiotics mentioned earlier, the tetracyclines are not bactericidal (they fail to kill sensitive bacteria) but are bacteriostatic and prevent sensitive bacteria from proliferating. This is rarely a practical limitation, unless the patient also lacks immune defenses.

A major indication for use of a tetracycline is for exacerbations of chronic bronchitis where the broad-spectrum of activity particularly against *H. influenzae* is useful. Otherwise tetracyclines are valuable for a common chest infection known as primary atypical (mycoplasmal) pneumonia and a variety of exotic infections, such as trachoma (a major cause of blindness in some African countries and the Middle East), cholera, and typhus. Tetracyclines may also be used for the treatment of syphilis or gonorrhoea in penicillin-allergic patients and have the advantage that they are also effective against non-specific urethritis, which is frequently associated. Tetracyclines are often highly effective for the treatment of acne, the common skin disease of adolescence.

The tetracyclines have a great variety of possible adverse effects but invariably cause irreversible staining of the developing teeth. They should not therefore be given to pregnant women or to children below the age of 12 years as there is no condition (in the Western world at least) for which there is not a suitable alternative antibiotic. More seriously, tetracyclines (apart from doxycycline and minocycline) can aggravate renal failure and can cause liver damage if dosage is excessive. The fungal infection, thrush, probably more commonly follows treatment with tetracyclines than any other antibiotic. Nevertheless, tetracyclines are widely used because these adverse effects are overall uncommon if appropriate precautions are taken and the broad-spectrum of activity is a useful 'blunderbuss' treatment if no bacteriological diagnosis is available.

Examples of tetracyclines are tetracycline itself, oxytetracycline, doxycycline (long-acting) and minocycline. These drugs are usually taken orally and, apart from the last two, are prone to cause nausea and sometimes diarrhoea.

Erythromycin

Erythromycin has a totally different chemical structure from penicillin but has a very similar spectrum of antibacterial activity. It is a useful alternative for penicillin-allergic patients. It is also effective in mycoplasmal pneumonia and for some penicillin-resistant staphylococci. It is often given for sinusitis, and for the prophylaxis of whooping cough and diphtheria when there are outbreaks and many children are at risk. An important newer use for erythromycin is for the treatment of legionnaires' disease, though frequently such patients are elderly and too debilitated to benefit by the time treatment is started.

Erythromycin is generally speaking remarkably harmless and one of the safest of antibiotics, but it is somewhat erratically absorbed and is generally less potent than penicillin.

Streptomycin and related drugs

Streptomycin was the first effective antituberculous drug. It has been life-saving for many victims of this disease and for which it is virtually exclusively used. This group of drugs is effective against many dangerous bacteria but use is limited by their toxic effects. All are, in varying degree, capable of damaging the organs of hearing or balance, especially in overdose and in the elderly: they can also damage the kidneys but are nevertheless valuable for serious renal infections if doses are strictly regulated. All of these drugs have to be given by injection.

Other members of this group are kanamycin, tobramycin, and gentamicin. The last is less toxic and most widely used but netilmicin is claimed to be less toxic still.

Other antituberculosis drugs

As described in Chapter 8, several (three or even four) antituberculous drugs are given in combination initially, as the tubercle bacillus readily becomes resistant to any one of them. Such drugs (which are chemically quite different from the streptomycin group) include isoniazid, ethambutol, rifampicin and, as reserves, capreomycin, pyrazinamide, and cycloserine.

Pyrazinamide is particularly useful for tuberculous meningitis as it readily enters the cerebrospinal fluid. Isoniazid is also used for the prophylaxis of tuberculosis in those who are exposed to the infection. It can have a variety of toxic effects, including neuritis (nerve damage), but can also have a mood-elevating effect. This serendipitous phenomenon was noticed and made use of in developing the first effective antidepressive drugs, the mono-amine oxidase inhibitors. Rifampicin is used primarily for the treatment of tuberculosis but is effective against many other bacteria. There has been reluctance to use rifampicin for anything other than tuberculosis in case too many resistant bacteria start to proliferate. Toxic effects of rifampicin include a flu-like illness, nausea, vomiting or diarrhoea, shortness of breath (dyspnoea), renal failure, collapse or jaundice, and rashes. It also has the peculiar property of causing the saliva and urine to become alarmingly orange-red.

Clindamycin

This antibiotic can cause the kind of problem which unexpectedly gives a useful drug a bad name. Clindamycin is remarkably well absorbed when given by mouth, seems almost incapable of causing allergic reactions, and is useful against many common infections including anaerobic infections. Unlike penicillin, it diffuses well into tissues which have a poor blood supply and is valuable for bone infections.

Diarrhoea, usually mild, is clindamycin's only common adverse effect. Unfortunately, because of its activity against anaerobes, clindamycin destroys many of them in the gut. This allows a relatively resistant one (*Clostridium difficile*) to proliferate and produce a toxin which damages the bowel wall causing a rare condition known as pseudomembranous colitis, occasionally with fatal results. This complication is treatable by giving vancomycin or metronidazole by mouth. Nevertheless it is so dangerous that clindamycin is now reserved for infections for which there is no safer alternative. In brief courses of treatment, clindamycin is probably less dangerous than many believe, and in fact other antibiotics can occasionally cause pseudomembranous colitis.

Metronidazole

This widely used antibacterial was introduced for the common genital infection, trichomoniasis, but was found to be effective against many anaerobic microbes which cause gynaecological infections, peritonitis, trench mouth (Vincent's infection), and pseudomembranous colitis.

Metronidazole is effective only against anaerobic bacteria and protozoa and rarely causes significant toxic effects. Although there is concern about its use during pregnancy, there is no real evidence that it damages the fetus. One side effect of metronidazole is that if the patient takes alcohol during treatment, the interaction between these two causes flushing, palpitations, and a fall in blood pressure. These symptoms are sufficiently unpleasant for metronidazole to have been used for the treatment of alcoholism.

It should be said incidentally, that anaerobes are bacteria which can only live in the absence of oxygen. This may be made possible often by the presence of other bacteria which use up the oxygen. Anaerobes are common in sites such as gum pockets round the teeth and in the lower gut and can cause serious infections, especially if they reach other tissues. They typically produce foul smells—the odour of rotting carcases is an example. Possibly the most famous (or infamous) anaerobic infection is gas gangrene, which caused many deaths in the 1914–18 war. The importance of anaerobes in many infections was not appreciated until relatively recent years when better methods of isolating these bacteria (which are usually in mixed infections) were developed.

Tinidazole is similar to metronidazole but has a longer duration of action, allowing less frequent administration.

The sulphonamides and trimethoprim

As mentioned earlier, the sulphonamides were the first effective antibacterials. They are now mainly useful for urinary infections as they are excreted through the kidneys. They frequently cause minor toxic effects, such as rashes, but can damage the kidney if fluid intake is inadequate and may occasionally damage the bone marrow causing agranulocytosis.

Trimethoprim can enhance the activity of sulphonamides and the combination of sulphamethoxazole and trimethoprim, known as co-trimoxazole (Bactrim, Septrin), has become a widely used treatment for exacerbations of chronic bronchitis, urinary and severe gut infections such as some cases of typhoid fever. Co-trimoxazole has recently been shown to be an effective prophylactic against food poisoning in high-risk areas such as Mexico. It is also valuable for the pneumonia caused by the parasite *Pneumocystis carinii* which only affects immuno-deficient patients, particularly those with AIDS. However, AIDS patients have to be given very high doses, severe toxic effects are common and even with such treatment the infection is frequently fatal.

For many other infections trimethoprim may be as effective as co-trimoxazole and causes fewer toxic effects than the sulphonamide component.

Chloramphenicol

This is another drug where the toxic effects, even though uncommon, have outweighed its usefulness. It is active against many kinds of bacteria, and unlike penicillin readily gets into the cerebrospinal fluid surrounding the brain. Chloramphenicol can, however, damage nerves and cause rashes, vomiting, and diarrhoea; worst of all, it can cause unpredictable and irreversibly fatal bone marrow poisoning. The danger is such that chloramphenicol is (or should never be) given except for life-threatening infections for which no other antibiotic is effective, such as some cases of typhoid fever. As an example of how antibiotics are abused, one of us can remember, many years ago, being given chloramphenicol for what was probably no more than a severe cold and despite the fact that it is totally ineffective against viruses.

Vancomycin

This rarely used antibiotic, is another example of a highly toxic drug which may nevertheless have to be used for life-threatening infections such as some of those affecting the heart (infective endocarditis) or those caused by staphylococci resistant to all other antibiotics. Vancomycin can damage the organs of hearing and the kidneys, can cause allergic reactions, and is so irritant to veins that it must be given in a dilute infusion extending over an hour. When given by mouth for the treatment of pseudomembranous colitis, however, vancomycin causes few problems.

Ciprofloxacin

This antibacterial has only recently been introduced. It is an example of the ingenuity of pharmaceutical chemists in that its mode of action is totally different from that of any other antibiotic. Ciprofloxacin acts by blocking the action of a bacterial enzyme called DNA gyrase which is responsible for the tight coiling of bacterial DNA. If this enzyme cannot work the bacterial DNA becomes uncoiled and can no longer be contained within the bacterial cell.

Ciprofloxacin has been shown to control many bacterial infections in immunodeficient patients, but its overall value has not yet been assessed.

Antifungal drugs

Fungal infections (mycoses) are of two main types namely superficial and deep (systemic). The superficial mycoses are the dermatophyte (ringworm) infections of the skin (tinea infections) which are usually little more than a nuisance. The deep mycoses, which are rare in Britain and other temperate areas, invade the tissues and are often life-threatening. *Candida albicans* is an exception in that it can cause both superficial infections (notably thrush in babies and in AIDS), but can also cause deep infections, particularly of the heart.

The distinction between the superficial and deep mycoses is important because the causative fungi in each group tend to respond to different types of antifungal drug. However, the newer antifungals (imidazoles) can be used for either type of mycosis.

Many tinea infections respond to local applications. Benzoic acid ointment (Whitfield's oinment) is effective for athlete's foot (tinea pedis) and for ringworm in other sites. This is one of the few examples of an infection which responds to an antiseptic. More resistant infections are likely to respond to clotrimazole (Canesten) ointment or one of the other imidazoles.

Troublesome tinea infections, particularly of the scalp or nails, are likely to need systemic treatment. Griseofulvin is taken by mouth but is then concentrated in the superficial (keratin) layers of the skin. It sometimes causes nausea, headaches, or rashes. An alternative is ketoconazole, but though it is effective, it can occasionally damage the liver.

Nystatin and amphotericin were the first effective antifungal antibiotics. They have a peculiar mode of action in that they perforate the cell membranes of the fungi, causing them to leak essential components and so die. Neither is significantly absorbed from the gut and so they have minimal toxic effects when given by mouth. However, they are only then effective for thrush if dissolved in the mouth as tablets, pastilles, or mixtures.

Amphotericin can be given intravenously for serious fungal infections (such as histoplasmosis, coccidiodomycosis, or cryptococcosis) which are frequently a consequence of AIDS. It remains the most potent of the antifungal drugs for a wide variety of these potentially lethal infections. However, it is very toxic when given by injection and allergic reactions or damage to the kidneys, among other complications, may limit its use.

Imidazoles such as ketoconazole are valuable for systemic treatment of many of the serious fungal infections. Though less toxic than amphotericin, ketoconazole very occasionally causes serious liver damage.

Genital candidosis (thrush), one of the most common sexually transmitted diseases, can be treated with preparations (creams or pessaries) containing nystatin, clotrimazole, econazole, miconazole, or similar drugs.

Antiviral drugs

Since viruses pirate the activities of the victim's own cells to reproduce themselves, it is particularly difficult to develop drugs which attack viruses without also harming the patient. There are, therefore, still only a very few effective antiviral drugs.

One of the most recent and most effective is acyclovir (Zovirax) which makes use of normally undetectable differences between host and viral enzymes with the result that it deceives the virus into making faulty DNA and thus stops it replicating. Acyclovir is also of remarkably low toxicity. It is most effective against herpes simplex and zoster infections, both of which are potentially lethal in immunodeficient patients. The efficacy and safety of acyclovir is such that it is now commonly given prophylactically to these vulnerable patients, with the result that their incidence and mortality have been significantly reduced.

Herpes simplex virus is the most common cause of brain infection (encephalitis) in temperate climates such as Britain. Though rare it is commonly fatal and those that survive are frequently left seriously disabled. Diagnosis, before brain damage is too far advanced, is difficult because the early symptoms resemble drunkenness. Until recently the diagnosis could only be confirmed by biopsy (taking a small specimen) of the brain—something not eagerly welcomed by everyone. However, it is becoming increasingly common to give acyclovir on suspicion as it may prevent serious brain damage at this stage, and if the clinical diagnosis turns out to be wrong the acyclovir is unlikely to have done any harm. This is not the ideal way to treat disease, but desperate conditions may call for desperate remedies.

Acyclovir can be used as a cream for cold sores of the lip, while for genital herpes recurrences can be suppressed by prolonged courses of acyclovir by mouth, though this is not generally recommended. For severe herpetic infections, such as herpes zoster (shingles) which is disablingly painful and debilitating (particularly in the elderly), acyclovir should be given in large doses by mouth or intravenously.

Acyclovir can cause rashes, gut upsets, and disturbances of liver or kidney function, but these toxic effects are rarely troublesome. Even those with pre-existing kidney disease who also have serious herpetic infections can be given acyclovir, but in smaller doses since less is lost in the urine.

Idoxuridine is an alternative antiherpetic drug which is effective for cold sores of the lip if applied early enough and moderately effective for herpetic eye infections in the form of eye drops, but is too toxic for systemic use.

Amantadine is an example of the extreme selectivity of action of some drugs in that it is effective in the prophylaxis of influenza but only for one strain (the A strain) of the causative virus. It may therefore be useful for vulnerable patients, particularly the elderly and children, when there is an outbreak of influenza caused by the A strain of this virus. However, amantadine is more frequently used for a quite different purpose that has nothing to do with its antiviral activity; it is moderately effective for Parkinson's disease and can lessen rigidity and tremor but increase normal muscular activity.

Vidarabine is another drug effective against herpetic infections but can only be given by very slow intravenous infusions of a dilute solution. Apart from this limitation, vidarabine is more likely to cause toxic effects than acyclovir.

Zidovudine (Retrovir; formerly known as azidothymidine or AZT) is active against the AIDS virus. It is clinically beneficial in the short-term, but it has serious toxic effects and its long-term value is uncertain. At present there is no better alternative treatment but the mere fact that zidovudine has been developed carries with it the possibility that other more effective drugs can be produced.

Interferons

These substances are produced by the victim when attacked by a viral infection and act by blocking receptors (on the cell surface) which instruct the cell to make viral proteins. Unlike the previously described antiviral drugs which are highly specific, the interferons have this blocking effect on all viruses. Interferons are species-specific in that only interferons produced by human cells are effective in human viral infections. As a result, it has proved difficult to produce interferons in useful quantities. Although interferons were discovered over 30 years ago (and at that time thought to be the miracle cure for viral infections), they have only recently been put into production and licensed for use for a few specific conditions.

Experimentally, interferon nasal sprays appear to reduce the incidence of colds in those chronically exposed to infection and it is thought to be beneficial for severe viral infections such as Lassa fever. In practice the only proven use for an interferon is that of alpha-interferon (Wellferon) in the treatment of a rare form of leukaemia (hairy cell leukaemia) which may be caused by a virus. Another interferon (Roferon-A) has been licensed for the treatment of AIDS-related tumours, and in particular Kaposi's sarcoma, but its value is uncertain.

The toxic effects of the interferons are severe and include a prolonged, feverish, flu-like illness with lethargy, depression, and difficulty in sleeping. One patient treated with an interferon said after three months of these symptoms that he would rather have died from his infection.

In summary, it can be said that antibacterial drugs represent enormous advances that, within certain limitations, have revolutionized the treatment of infections. In a way, pharmaceutical science has been almost too successful in that many expect a cure for every disease. A current problem is myalgic encephalomyelitis which is characterized by headache, fever, myalgia, muscular weakness, and depression. Those who suffer from it are anxious to have it confirmed as a 'real' disease and, in particular, an infection since they feel this will lead to a cure. Many aspects of the disease suggest that a virus may be responsible, but even if this is established beyond doubt, the unfortunate patients are no better off until an appropriate antiviral drug is developed—a possibility but by no means a certainty. Non-specific antiviral treatment with interferon is unlikely to help. Its effects are very similar to the symptoms of myalgic encephalomyelitis and it is just possible that if a virus is responsible for this disease, its effects are at least partly mediated by the patient's own production of interferon. Viruses, notably some strains of influenza virus, can cause changes in the brain which result in severe depression. One of the symptoms of myalgic encephalomyelitis is depression and this too could be caused by a virus. Treatment of the depression (Chapter 4) may be easier than eliminating the virus and some sufferers have been greatly helped in this way. Sadly, a diagnosis is really no more than giving a name to a disease—it does not automatically help the patient. The fact that there is as yet no answer to the problems of the mysteries of myalgic encephalomyelitis illustrates the often unpalatable fact that, despite all endeavours, there are finite limits to medical treatment as there are to every other aspect of human knowledge.

4

The brain and nervous system: neurological, psychological, and emotional disorders.

All physical and mental activity depends on the nervous system which comprises the brain, spinal cord, and peripheral nerves which spread throughout the body. One major part of the nervous system transmits the impulses from the brain which initiate muscular activity (the motor system), while another transmits pain, sight, hearing, and other sensations.

Neurological disease refers to physical disorders of the nervous system. Nevertheless, it is probable that many psychiatric or emotional disorders, such as depression and anxiety states, result from subtle biochemical changes within the brain. Certainly, there are drugs which induce depression or anxiety, while others can relieve them. Depression, as discussed later, is just as much a disease as many others and it is unfortunate that the ignorant, and even some of those that suffer from it, regard such illnesses as a sign of weakness and merely due to nerves. Abnormal feelings depend on activity within the nervous system, but the terms 'nerves' or 'neuroticism' should certainly not be used in any pejorative sense, or as a criticism of the sufferers from such complaints.

Neurological disorders

Epilepsy

It is a common misapprehension that epilepsy is a sign of a mental defect; this is quite untrue except insofar as both epilepsy and mental defects can result from severe brain damage. Epilepsy, however, can be socially embarrassing and dangerous in a pedestrian or motorist if consciousness or concentration is lost even momentarily in present day traffic.

In most patients who have fits, no brain abnormality is found. This is termed primary or idiopathic epilepsy—a slightly pompous way of saying that we do not know the cause.

Major epilepsy (tonic–clonic fits), is synonymous with convulsions. It is not a single disease, but a sign of many conditions which can affect the brain. These include scars, congenital abnormalities, obstruction of blood vessels, and brain tumours. Very high blood pressure and some chemical abnormalities of the blood—which have secondary effects on the brain— can also cause fits.

Epilepsy consists of discrete attacks which last from seconds to minutes. The form and severity of attacks vary greatly and range from barely noticeable lapses in attention ('absences') to dramatic episodes of loss of consciousness associated with violent jerking of the limbs and pallor or blueness of the skin. These attacks appear to result from sudden abnormal bursts of electrical activity in part of the brain, anticonvulsant drugs aim to minimize or stop these discharges.

In assessing a patient who appears to suffer from epilepsy several decisions have to be made. First, whether the episodes are true (primary) epilepsy or if they are some other condition such as faints. The next essential is a detailed investigation to find out if the fits have a specific cause such as a low blood sugar (hypoglycaemia), low blood calcium (hypocalcaemia), brain tumour, or brain abscess, which has to be treated. Investigations frequently include radiographs of skull, electroencephalography (EEC) (to record patterns of electrical activity in the brain), brain scans, and examinations of cerebrospinal fluid removed by lumbar puncture.

Even in patients with primary epilepsy, fits can be triggered by various factors, for example, lack of sleep, frustration and boredom, alcohol (particularly during the hangover phase), flashing lights (faulty fluorescent tubes, television, and disco lights). Drinking excessive amounts of water (which also disturbs brain function) can be responsible in some patients. Thus for each person with epilepsy triggers must be looked for and, if possible, avoided. Drug treatment is started only if these measures fail. The choice of drug depends on the type of epilepsy.

Petit mal

This kind of epilepsy only affects children and stops before adolescence. During attacks, often also termed absence seizures, the child loses concentration for a few seconds and the eyes turn upwards. These episodes are brief, but there may be as many as several hundred in a day. Diagnosis is helped by a characteristic pattern of the EEG.

Drug treatment is frequently effective in either abolishing or greatly reducing the number of fits. Ethosuximide (Zarontin), for example, can abolish petit mal attacks and is only used for them as it has no effect on other forms of epilepsy. The way this drug is given illustrates many of the principles of drug treatment of epilepsy in general. The action of ethosuximide lasts for over 24 hours and so it needs only to be given once a day— the late evening is usually convenient. It may cause mild indigestion and sleepiness, but these are unlikely to be troublesome during the night. The starting dose is very small but, if necessary, is gradually increased. A diary of fits and general symptoms is kept by the patient (or parent). After a week at each dose level, the dosage may be kept unchanged if all fits have stopped, or increased if fits continue. If, however, adverse effects appear, the physician may decide to stop the drug; this must be done by a gradual reduction in dose.

If fits are not controlled or adverse effects are unacceptable, alternative drugs such as sodium valproate (Epilim) or clonazepam (Rivotril) can be tried. It is simplest and safest to use a single drug for epilepsy, and in most cases is adequate. In changing from one drug to another it is important to avoid any sudden alteration in dose. In particular, no anticonvulsant drug being given at moderate or high doses should be stopped suddenly. This can result in a dangerously severe resurgence of epilepsy. Thus, if one drug has to be stopped and another started, the dose of the first is gradually reduced whilst at the same time increasing the dose of the second drug. Therefore, when one drug has been finally stopped, the replacement is being given in substantial doses.

Sodium valproate can cause indigestion and is mildly sedating. An unusual feature is that it can cause some loss of hair, but the hair grows again (often curly if it was straight before) when the drug is stopped. Clonazepam is a close relative of diazepam (Valium), and can have similar side-effects, in particular tiredness and sleepiness.

Treatment of petit mal is continued for several years and usually maintained until adolescence, when it is gradually tailed off over three or four months. If fits recur, the dose is raised again.

Major epilepsy (tonic–clonic epilepsy: grand mal)

These fits can develop at any age, and the convulsions can be dramatic. The victim experiences a warning (aura) that an attack is imminent. This may be a vague sensation in the pit of the stomach or a precise set of symptoms which relates to a particular part of the brain. For example, the temporal lobe is concerned with the perception of smell, memory, and time, and with expression of emotions. The aura preceding fits due to an abnormality in the temporal lobe of the brain may therefore be an hallucination of a particular smell, a memory of a past event, or the experience of an emotion such as fear.* Immediately after the aura, the patient loses consciousness and falls down. The muscles contract powerfully (tonic phase) and then move violently and jerkily (clonic phase). The patient, being unable to breathe properly, becomes blue and pale due to lack of oxygen. After a few minutes consciousness returns, but complete recovery may take several hours. During this phase the patient may want to sleep.

There is no treatment for an epileptic attack once it has started. All that can be done is to protect the patient from injury, either through falling or as a result of the violent muscle spasms. Treatment is therefore aimed at preventing attacks.

The commonly used drugs for epilepsy are carbamazepine (Tegretol), phenytoin (Epanutin), and sodium valproate (Epilim). As always, the decision to use drugs should be based on a consideration of their benefits versus their risks. No drug is completely safe, but the physical danger of fits and the social ostracism they may cause, have to be weighed against the hazards of drug toxicity. Certainly two or more fits each year amount to a significant handicap.

If treatment is completely successful, the same drug is continued for about three years before the dose is slowly and cautiously reduced to see whether it is possible to stop treatment altogether without recurrence of fits. As already mentioned, drugs should not be stopped suddenly. This can cause repeated fits to follow one another without intermission (status epilepticus). Unlike a single fit, it is essential to control status epilepticus as quickly as possible because the resulting asphyxia can cause brain damage or death. For this purpose, an injection of diazepam (Valium) or clonazepam is usually quickly effective, but expert hospital care is imperative.

Carbamazepine is effective in tonic–clonic fits, but has no action on petit mal. It frequently produces minor toxic effects,

---

*Pure temporal lobe epilepsy is incidentally an unusual form of the disease and the strange sensations and visions seen by the nurse in *The Turn of the Screw* by Henry James, are a vivid and accurate description of how it may effect some patients.

such as giddiness, unsteadiness, and sedation. Serious toxicity is very rare but the drug stimulates the liver to destroy other drugs and hormones quickly. Thus, if a woman taking an oral contraceptive is given a course of carbamazepine, pregnancy may result owing to diminution of the action of the pill.

Phenytoin is a long-established drug; this is an advantage because its full potential for producing toxic effects is better known. It is as effective as carbamazepine in tonic–clonic epilepsy but, rarely, can interfere with blood formation and cause anaemia or very occasionally more serious blood disease. Much more common and less serious side-effects are swelling of the gums, thickening of the facial features, increased oiliness of the skin, worsening of acne, increased hairiness, and (if the dose is excessive) giddiness and unsteadiness. Phenytoin is of no value in the treatment of petit mal.

Phenytoin, and probably other anticonvulsant drugs including phenobarbitone, primidone, sodium valproate, and carbamazepine, may harm the unborn child. Epileptic women have two to three times the risk of giving birth to babies with congenital abnormalities, particularly harelip, cleft palate, and heart defects. It is likely that these lesions are at least partly due to taking anticonvulsant drugs. However, uncontrolled fits in a pregnant woman can also harm the fetus and may even cause fetal death.

Sodium valproate is a drug which is effective in a wide range of epilepsies. In tonic–clonic epilepsy it is as effective as carbamazepine and phenytoin but can very occasionally damage the liver, as well as having other less severe side effects.

Phenobarbitone (Luminal) and primidone (Mysoline) are similar drugs. They are effective only for major epilepsy, but only used if other drugs have been found to be unsatisfactory. Their main disadvantage is that when given in sufficient amounts to be effective, they usually cause drowsiness. They are also even more prone than other anticonvulsants to cause a serious rebound increase in fits on suddenly stopping their administration. Like carbamazepine and phenytoin, they stimulate the liver to eliminate some drugs and hormones and thus may make the contraceptive pill ineffective.

Parkinson's disease

Parkinson's disease is a condition which particularly affects the elderly; it consists of tremor (hence the old term 'shaking palsy'), stiffness of all the muscles, and difficulty in starting movements. The disability tends to be progressive and can be so severe that all useful movement becomes impossible and the patient is unable to walk, dress, or eat without assistance. The interference with movement, the slowing of speech, immobility of the face producing mask-like unresponsiveness, and sometimes drooling of saliva may, sadly, cause the patient to appear stupid.

Parkinson's disease is due to lack of a single chemical in the brain, called dopamine. Dopamine is made by nerve cells in a region of the brain which is concerned with smoothing out and co-ordinating muscular actions. In Parkinson's disease these particular brain cells degenerate and disappear.* Thus, dopamine is no longer produced and is not available to control the quality of muscular movements.

Drug treatment usually improves the quality of life and prolongs life expectancy of patients with Parkinson's disease. The drugs used are: levodopa (Brocadopa; Larodopa and also in Sinemet and Madopar), bromocriptine (Parlodel), amantadine (Symmetrel), selegiline (Eldepryl), and the benzhexol group.

Levodopa is given by mouth but some gets into the brain and is there changed into dopamine. Dopamine has a direct, beneficial action in Parkinson's disease by correcting its deficiency in the

---

*One of the ironies of present-day drug trafficking is that a by-product (MPTP) of an illicit synthetic drug of addiction kills off, with astonishing precision, the particular brain cells whose destruction is responsible for parkinsonism. MPTP is so potent in this action that the victims of its use become frozen into total muscular rigidity and may appear to be unconscious. The unravelling of the complex story of MPTP is likely to throw considerable light on the mechanism and control of Parkinson's disease.

brain (the underlying cause of the illness) as already described. In practice three-quarters of patients with this condition are greatly helped. Those who would otherwise have to give up work or looking after a family can often resume normal activity. It is remarkable that swallowing a chemical which gets all over the body can supply dopamine to just the small and remote part of the brain where it is needed and can correct the deficiency resulting from loss of brain cells.

Much of the swallowed levodopa is wasted because it is converted to dopamine in tissues other than the brain. Moreover, dopamine can also cause nausea, vomiting, and irregularities of the rhythm of the heart. The addition of another agent which prevents the conversion of levodopa to dopamine outside the brain allows a much smaller dose of the drug to be used, because less is wasted in non-brain tissues. The incidence of nausea and heart complications is thus greatly reduced. Agents which are given with levodopa for this purpose are carbidopa or benserazide, and it is a combination of levodopa and either of these which is the most effective drug treatment for Parkinson's disease at present.

Nevertheless, the disease is progressive and treatment in some patients becomes less effective over the years. As a consequence, increasing doses of levodopa may be needed and may lead to toxic effects. At its peak, the effect of a high dose can be involuntary jerking or writhing movements of the limbs, screwing up of the face, and twisting of the trunk resulting from high concentrations of dopamine formed in the brain. This is followed by a short period of relief from symptoms and then, as the effect of the drug wears off, the stiffness and immobility returns. Other consequences of long-term treatment with high doses are psychological disturbances. These include vivid nightmares and depression. Not all patients suffer toxic effects—about half continue to benefit from treatment after five or more years, but a quarter have to stop treatment with levodopa because of toxic effects.

Bromocriptine is a drug manufactured from the rust fungus of cereals (*Claviceps purpurea*). It acts on the brain in a very similar way to dopamine and shares the same benefits and toxic actions as levodopa. The only difference is that bromocriptine has a longer action (about 6–8 hours) than levodopa (2–6 hours) and so is less prone to produce rapid swings from involuntary movements (due to excessive drug effects) to rigidity due to the drug action wearing off.

Amantadine was originally developed as an antiviral drug but was accidentally found to benefit patients with Parkinson's disease. It works by making more dopamine available to the brain—probably by facilitating its release from stores in surviving nerve cells. Not so many patients are helped as by levodopa, but some do very well with amantadine. One limitation, however, is that in 30–40 per cent of patients the improvement is not maintained beyond two months. Serious side-effects are not common, but amantadine can cause confusion in the elderly.

Selegiline is an agent reserved for patients with advanced disease, in whom levodopa has to be given in over-large doses. Selegiline acts by blocking the enzymes in the brain which normally destroy dopamine. It is not given on its own but is added to a regimen of levodopa plus benserazide or carbidopa. The side-effects of selegiline are usually mental stimulation and resemble the actions of caffeine (the same effects as drinking strong tea or coffee). Thus, insomnia or a feeling of excitement may result.

The benzhexol group of drugs includes benzhexol (Artane), benztropine (Cogentin), and orphenadrine (Disipal). They are less strikingly effective, but are often used in early disease, in which they can lessen pain and disability due to muscular rigidity. They can also be used in more advanced disease in combination with levodopa. These drugs act by antagonizing the effects of one of the chemical messengers in the brain—acetylcholine. This is a stimulatory substance for certain brain cells and its action is, in effect, the opposite to that of dopamine. Excessive amounts of acetylcholine aggravate Parkinson's disease. The ill-effects of this group of drugs include dry mouth, constipation, and difficulty in urination. Some forms of glaucoma can be worsened, and confusion sometimes results in the elderly.

Psychological and emotional disorders

Psychiatry is concerned with any aspect of human behaviour, thought, and communication which can cause distress. Everyone experiences situations which cause fear, worry, or sadness. These are all psychological phenomena which, if sufficiently intense, could induce us to look for help. Such help may be by talking about the experience to a relative or friend or to a professional counsellor such as a priest or psychiatrist. Discussion of one's problems is a powerful and effective medicine—the person with whom the conversation takes place is a matter of one's cultural background, past experiences, opportunities, and expectations. Because this book is mainly concerned with traditional medicine and medicines, we will use psychiatric terminology and ideas. However this is not to imply that psychiatrists have a more accurate insight into the human condition than say a painter, poet, musician, novelist, clergyman, or mystic. In medical terms, the major types of mental illness are:

Anxiety states

Affective illness (depression and mania)

Schizophrenia

Dementia

Anxiety states

Anxiety is fear and fear is one of the most important emotions to have evolved in man and other animals because it triggers the appropriate responses to danger. To survive, it is necessary to learn which situations are hazardous and how to cope with them. The reactions have been epitomized as fright, fight, and flight. The physical events in the body which accompany these reactions are an increase in the rate and force of the heartbeat, inhibition of digestion, increased blood flow through muscles, tremor, dry mouth, and sweating. Although most of these would be useful in a fight, or in running away from one, a dry mouth, tremor and sweating may not help some everyday stresses such as facing an interview, having a driving test, or making a speech.

The mechanism for all of these responses depends on activation of emotional centres (limbic system) in the brain. This causes activation of peripheral nerves to the heart, skin, and other organs which are not under direct voluntary control but are controlled by the autonomic nervous system. One response is secretion of adrenaline and noradrenaline into the blood from the adrenal glands. Adrenaline released in this way travels to all parts of the body and increases the amount of energy available to the heart, muscles, and brain by causing the breakdown of body food stores to glucose. Such frantic preparation for stress may not necessarily be appropriate. Nevertheless such diverse activities as teaching, selling, singing, and sports do seem to go better if there is the added edge of emotional involvement. Many actors find that a little stage-fright is helpful to their performance. However, there is a limit to the amount of tension and anxiety which is desirable. Drying of the mouth can disturb an actor's performance and tremor can be disastrous for a violinist's bowing arm. Extreme terror can even prevent someone from running away (frozen on the spot with fright). Nevertheless all these are examples of understandable fear.

Anxiety neurosis is the situation in which fear is experienced, but the stimulus which caused it is not appropriate. For example, it may be daunting to have to make a speech at a wedding, but it seems unreasonable to be terrified of going to a supermarket and having to face the person at the checkout point. Even less understandable is fear experienced for prolonged periods without any external threat at all. And yet such irrational fears are amongst the most common psychological disturbances—at least one-third of the population experience an anxiety state of sufficient severity to require some sort of professional help.

Anxiety states can arise without any apparent cause, or may start at the time of a life event which is obviously stressful (such as the death of a close relative) or an event which is not perceived as a stress (such as marriage, retirement, or the children leaving home). In addition, physical diseases can be accompanied by anxiety symptoms.

The manifestations of an anxiety state may be very varied. In some people the sensation is obviously one of anxiety, uneasiness, mental tension, or even panic. The psychological symptoms may not be so obviously based in anxiety but may be irritability, difficulty in falling asleep, inability to concentrate, constant feelings of tiredness, or the inability to enjoy company. Even more difficult to connect with anxiety is when the condition presents as physical symptoms which might suggest organic illness. Thus, the patient may experience chest pains, pains and stiffness in joints or muscles, headache, blurring of vision, pins and needles in the hands and feet, rapid or sighing breathing, the sensation of having to take a deep breath, bowel disturbances, bloating of the abdomen, impotence, frigidity, or a frequent need to urinate. Apart from these symptoms producing general problems of diagnosis, even if anxiety is strongly suspected, it may be important to exclude other kinds of underlying illness.

An important aspect of management of anxiety is establishing a reasonably firm diagnosis. Management of the condition is difficult, and because of this it is tempting for the patient and doctor to think in terms of drug treatment. Medicines are easy to take and may be very effective initially in suppressing tension and anxiety. However, the benefit is short-lived and on stopping the drug, a rebound increase in anxiety may be experienced.

Anti-anxiety drugs (anxiolytics; minor tranquillizers)

Drugs, particularly alcohol, which reduce feelings of anxiety have been available to man throughout history. Since the nineteenth century, synthetic drugs, such as the barbiturates, have been introduced and these have actions which partly overlap those of alcohol. Barbiturates such as phenobarbitone, like alcohol, can diminish emotional tension and anxiety. However, these benefits are attained at the cost of sleepiness, impairment of mental function, and unco-ordinated movement. Large doses impair an even wider range of brain function—death can be caused by inhibition of the nerve centres for breathing.

In 1961 a new group of drugs, the benzodiazepines, was introduced. The first was chlordiazepoxide (Librium) and this has been followed by diazepam (Valium), nitrazepam (Mogadon), flurazepam (Dalmane), and many others. All these drugs have similar actions: they lessen anxiety, and induce sleepiness and muscular relaxation: they also suppress epilepsy. The remarkable ability of the benzodiazepines to reduce feelings of fear has led to research into their mechanism of action. They appear to act by increasing the effects of a natural chemical messenger in the brain called gamma-aminobutyric acid (GABA).

The benzodiazepines were of particular interest because anxiety and emotional tension could be reduced with less sedation and loss of co-ordination than that produced by alcohol or phenobarbitone. However, sufficiently large doses of benzo-diazepines cause sleepiness and even with small doses the risk of motor vehicle accidents is increased. The overall inhibition of brain activity is also less than that produced by alcohol or the older tranquillizers, in that the centres controlling breathing and blood circulation are not seriously disturbed in healthy individuals. However if another sedative drug, such as alcohol, is taken at the same time as the benzodiazepine, or if the subject suffers from lung disease such as bronchitis or asthma, then dangerous depression of respiration can result.

The benzodiazepines would seem to be very useful drugs in lessening the anxiety and tension of many aspects of human suffering. However, when they are used to treat long-lasting anxiety, serious problems arise. The first is that tolerance develops, so that even though the drug is taken every day anxiety returns after one or two weeks of benefit. Increasing the dose of the drug gives only temporary relief and side-effects may worsen. Stopping the drug because the benefit from it has waned can result in a withdrawal state and this is the most troublesome side-effect. In other words, addiction (or dependence) develops. The withdrawal features are typically insomnia, loss of appetite, increased mental and physical tension, and attacks of panic and terror. In addition to muscular tension, the muscles may ache and pains may be experienced. Several days after stopping large doses of the drug given over a prolonged period, fits may start: if these are prolonged, life can be threatened.

Because of the dangers of benzodiazepines and the development of tolerance they are not recommended in the long-term management of anxiety. An important exception is in those individuals where some other condition may be aggravating the anxiety. Thus, some patients with depression also suffer from agitation and tension. Such people may respond to antidepressive therapy, although even this is no reason to impose drug treatment on anyone for more than a few months. Sometimes anxiety and nervousness are symptoms of thyroid disease. This can be treated well with appropriate drugs which act on the thyroid gland. Even in patients with anxiety without an underlying or aggravating cause, if the main distress is due to awareness of a rapid and forceful beating of the heart (palpitations) or due to tremor, these can be reduced by drugs which mainly act on the periphery of the body. These drugs belong to the beta-blocker group and include propranolol (Inderal), oxprenolol (Trasicor), atenolol (Tenormin), and metoprolol (Lopresor). An important use of benzodiazepines is in the control of anxiety caused by frightening procedures in hospital. For this purpose diazepam or a midazolam is given by intravenous injection.

Most people with long term or recurrent anxiety should not rely on drugs to relieve their symptoms. Psychotherapists have developed alternative methods which, in some patients, are more effective than drugs and are without any of their side-effects. Such techniques include relaxation therapy. This can be learned from a therapist, but a book or audiotape can also be used. It is not known why muscular relaxation and regular breathing are conducive to reduction in mental tension, but it does seem that anxiety provokes muscular tension, and that muscular tension reminds the subject of mental tension. It is possible that muscle relaxation acts by breaking this vicious circle.

Affective illnesses (disorders of mood)

Depression

Depression is abnormal sadness. The borderline between reasonable and unreasonable unhappiness is difficult to define. The death of a close relative or collapse of a personal relationship are common and natural reasons for feeling sad—even to the extent of prolonged grief. But sustained and profound grief following a minor threat to self-esteem or embarrassment suggests an over-vulnerable personality. In the past psychiatrists divided depression into two main groups:

1. Endogenous depression in which the pre-existing personality appeared to be robust and no obvious precipitating cause was present.
2. Reactive depression which was said to be more likely in vulnerable personalities and follow a psychologically stressful event.

The main difficulty with this classification is that the majority of depressed people show features of both groups. Also, if environmental stresses are severe and prolonged enough, then almost anyone can become depressed. Many of the survivors of the Nazi concentration camps were left with long-standing psychological difficulties, including depression. In view of the personality and environmental factors in depressive illness, it is at first surprising that there is any support for the idea that depression could have a simple chemical basis.

The activity of nerve cells in the brain is controlled by chemical messengers (transmitters). Some of these transmitters are, in chemical terms, amines and include noradrenaline, serotonin, and dopamine. It appears that drugs which lower amine levels in the brain cause depression, while drugs which raise the amine levels can relieve depression. Reserpine (as a crude plant extract) for example has been used in India for hundreds of years to treat abnormal restlessness and excitement. It lowers brain amine levels and though at one time was used to treat hypertension, causes unacceptable depression in many patients. The antidepressant drugs by contrast, increase the availability of amines in the brain. They include:

1. Tricyclic antidepressants, such as imipramine (Tofranil); amitriptyline (Triptafen; Tryptizol), and dothiepin (Prothiaden)
2. Mianserin (Norval; Bolvidon)
3. Trazodone (Molipaxin)
4. Monoamine oxidase inhibitors, for example phenelzine (Nardil)

The tricyclic antidepressants are the most widely used group. Amitriptyline, for example, will alleviate about 70 per cent of attacks of depression within 2–4 weeks. The action is not immediate and the patient should be warned of this. The drug induces fatigue and sleepiness in most people, and is therefore best given as a single dose at night before going to bed. If another sedative is taken at the same time, then this action is greatly increased and deep unconsciousness can result: alcohol (which is a sedative) can interact with amitriptyline in this way.

Tricyclic antidepressants can interfere with some of the functions of the body which are not under voluntary control. For example, the secretion of saliva is reduced, causing a dry mouth. There may be difficulty in emptying the bladder and bowel. Focusing the eyes on near objects can be impaired. Large doses are dangerous and can cause abnormalities of the heart rhythm and fits.

The tricyclic antidepressants improve the ability of brain nerve cells to make available noradrenaline and other amines. It may be that this is how they exert their benefit in depressive illness.

Mianserin was considered to be an advance in the treatment of depression because large doses do not alter the rhythm of the heart, and there is little or no disturbance to the eye, bladder, bowel, and body secretions (all of which are affected by the tricyclic antidepressants). The main toxic effects of mianserin are sleepiness (and therefore potential for an adverse reaction with alcohol) and a proneness to dizziness and fainting on standing up after sitting or lying. In the main, the early promise shown by this drug has been confirmed: accidental or intentional overdose does not carry the same dangers on the heart compared with the tricyclic antidepressants. However, rarely mianserin can cause liver damage (and lead to jaundice) or bone marrow damage (and cause dangerous reduction in the white cells in the blood).

Trazodone is another sedating antidepressant which is much safer than the tricyclic antidepressants. Serious toxicity to the heart, eye, bladder, and bowel has not been described.

Monoamine oxidase inhibitors such as phenelzine are dangerous drugs and therefore have to be used with care and under close supervision. Whereas all the previously mentioned antidepressants appear to act by conserving and redistributing pre-existing brain amines, the monoamine oxidase inhibitors actually increase the stores of these substances in the brain. If the theory that some types of depression are due to deficiency of amines in the brain is correct, then these drugs should relieve depression. They help about 70 per cent of patients suffering from an attack of depression.

The danger of this group of drugs is due to their ability to increase stores of noradrenaline in all parts of the body. Noradrenaline can raise the blood pressure if it is released into the circulation. Certain simple drugs and dietary substances have the ability to release noradrenaline from body stores. These include ephedrine and phenylethylamine (in nose drops and common cold 'cures') and tyramine (in cheese and fermented foods and drinks). If the stores of noradrenaline are very large then eating even a small amount of cheese or using the above types of drugs can dramatically raise the blood pressure and may lead to a stroke or sudden death.

Monoamine oxidase inhibitors are now mainly used by specialists in the treatment of depressive illness. They are reserved for those few patients whose illness has not responded to other forms of treatment. A patient taking these drugs should be given a card to carry at all times, stating which drug is being taken and which foods and medications must be avoided.

Electroconvulsive therapy (ECT)

All the antidepressive drugs mentioned, frequently take several weeks to work. This means a longer period of suffering and there may be a danger of suicide. ECT can produce relief from depression within a day or two and therefore can be used to alleviate depression during the period it takes for drug activity to start.

Mania and cyclothymia

Mania is abnormal elation, energy, and optimism in episodes interspersed with normal health or may develop in an individual who also experiences periods of depression. A phase of mania can be terminated with a drug of the major tranquillizer group such as chlorpromazine or haloperidol.

Recurrent attacks of mania or depression, or of both affecting the same patient at different times, may respond to lithium. Lithium is an element, similar in some ways to sodium. It is given as a salt (often lithium carbonate), which stabilizes mood and may prevent abnormal mood swings (called cyclothymia). How lithium acts is not known, but some current theories suggest that it stabilizes the response of brain cells to amine transmitters so that alterations in release of brain amines have less effect on mood. Lithium works satisfactorily if it is maintained at a critical concentration in the blood. Low concentrations are ineffective and high concentrations produce toxic effects which include tremor, dehydration, and diarrhoea. Because a particular dose will produce different blood concentrations in different people it is necessary to check blood levels repeatedly in order to adjust the dose correctly for each patient.

Schizophrenia

Schizophrenia is the most terrible of all non-fatal diseases. It is a shattering of the personality and a destruction of the links between the mind and reality. (Split personality is an unrelated condition— a rare form of hysteria in which the sufferer, at different times, exhibits two distinct personalities). Schizophrenia is common— one estimate is that it affects approximately 1 per cent of the population. The features of the disease include hallucinations which take the form of hearing voices, delusions of persecution and bizarre ideas (such as the belief that one's thoughts are being controlled or stopped by some external influence).

It was partly because of this illness that padded cells, strait-jackets and long-stay mental institutions were introduced. One of the great milestones in drug development was the appearance of agents which were effective against schizophrenia. The first drug of this type was chlorpromazine (Largactil). Since then many other drugs have been developed and used in schizophrenia. However, most of these have similar properties and are not greatly different from chlorpromazine, which is still extensively used in this disease.

The cause of schizophrenia is not known, but there is evidence that an abnormal increase in the actions of some chemical transmitters in the brain may be responsible. Drugs such as levodopa and amphetamine in large doses can produce delusions of persecution and hallucinations in normal people. These agents raise the activity of dopamine by different mechanisms, but the end result is the same, dopamine concentrations are increased in nerve cells in the brain. The parts of the brain which are relevant to these actions are involved in the learning and expression of emotional responses; these regions comprise the limbic system.

Chlorpromazine and similar drugs act by blocking the dopamine-sensitive (receptor) part of nerves. This may explain why they benefit patients suffering from schizophrenia. Such drugs are known as neuroleptics (major tranquillizers). As well as reducing or even abolishing the clinical features of schizophrenia, they are also effective in mania, delirium, and other excited states with serious mental disturbance.

The toxic effects of the neuroleptics are mainly due to widespread reduction of dopamine actions in the brain. Powerful inhibition of the limbic system results in emotional indifference (observed as a lack of emotional involvement in external events) or in depression. Dopamine is also a transmitter in those parts of the brain concerned with controlling the smoothness and steadiness of muscular movement. Disturbance of these parts by the neuroleptics can cause tremor and slow, stiff movements similar to those seen in patients with Parkinson's disease. When the drug is stopped, the tremor and stiffness disappear.

There is another abnormality of movement which can be caused by this group of drugs and is irreversible. Withdrawal of the neuroleptics may even worsen the abnormal movements, called tardive dyskinesia. These usually consist of repeated pouting, lip smacking, tongue protrusion, and chewing movements. The patient cannot voluntarily stop them and sometimes other parts of the body, such as the limbs, trunk, or breathing muscles, are involved. Tardive dyskinesia is a terrible price to pay for the help from the neuroleptics, and some surveys have shown that up to 30 per cent of patients on long-term drug treatment develop this condition. Much research effort is now focused on developing a neuroleptic which has an action on the limbic system, with less effect on the brain centres controlling movement. Some of the newer drugs, such as pimozide, partially reach this goal: the antischizophrenic action is powerful, but there is relatively little disturbance of muscular movement.

Another development in this field, is the use of long-acting injections of neuroleptics so that deterioration does not develop due to the patient forgetting to take the drug. An example is fluphenazine decanoate (Modecate) which is given at intervals of 2–4 weeks.

Migraine and other headaches

There are many causes of headache. If there has been a change in pattern or increase in severity of headaches then these should be investigated. However, the majority of headaches are related to tension or to migraine.

Tension headaches are associated with emotional tension or anxiety and can be accompanied by tension in the muscles of the front and back of the skull. The muscular contraction may be severe enough to cause pain. Analgesic drugs may be ineffective but the headaches can be relieved by relaxation exercises.

Migraine attacks are due to excessive dilatation of blood vessels in the skull or on the surface membranes of the brain. This causes the throbbing headache. It usually responds to an analgesic such as aspirin or paracetamol. Frequently nausea and vomiting accompany the attack. To combat this, anti-emetic drugs including metoclopramide, cyclizine, or perphenazine may be given with or preferably before the analgesic. The small proportion of migraine attacks which are not helped by this regimen may require ergotamine. Because excessive dosage of ergotamine can itself produce headache and vomiting, it is essential that the recommended dose is not exceeded. Further overdose can cause severe spasm of blood vessels with resulting tissue damage. If the attacks of migraine are frequent (more than twice a month) then many patients find a reduction in frequency by taking daily a beta-blocker or pizotifen. Pizotifen is also rather sedating and usually increases the appetite. Thus one of its disadvantages is weight gain.

The distinction between tension headaches and migraine is now thought to be less well defined than hitherto. Migraine attacks often come on during or after emotional stress or excitement and they may respond to relaxation therapy.

Sleeping difficulties

Disturbances of sleep are common and in addition many people have distorted ideas of how much sleep they need or indeed of how much sleep they have had. Drugs for insomnia (hypnotics) are therefore much in demand and widely—indeed too widely—used.

Causes of disturbed sleep, such as pain, anxiety, or depression, may require treatment but there are probably many who are poor sleepers, just as others have digestive troubles, for no discernible reason. Loss of sleep is probably much less important than is generally thought, though one would hope that those with highly responsible jobs such as aircraft pilots or air traffic controllers, were not bleary eyed and yawning their way through their working day.

Unfortunately drugs are not a satisfactory means of obtaining sleep. Barbiturates such as amylobarbitone (Amytal) or butobarbitone (Soneryl), though effective, readily lead to severe addiction and are therefore, like the major opiates, Controlled Drugs. Overdose can also be fatal. Benzodiazepines such as nitrazepam (Mogadon) or temazepam (Normison) are also effective hypnotics, probably as a consequence of their ability to relieve anxiety, but tolerance and dependence readily develop. Drugs such as these should be used only if absolutely essential and only then for a short period. Tricyclic antidepressants, particularly amitriptyline, are often strongly sedating and may sometimes be useful for intractable insomnia; their side-effects, such as hangover and dryness of the mouth are unpleasant but virtually prevent dependence from developing.

Far more satisfactory than drug treatment are simple measures, such as making sure that one is tired (preferably as a consequence of exercise) before seeking sleep and avoiding overeating, tea or coffee, and too much alcohol just before going to bed. A warm milky drink or a sandwich may be helpful, as may reading in bed or relaxation exercises with the help of a tape player. Fighting sleeplessness usually worsens the situation and it is far more effective to accept the inevitable, turn on the bedlight and start to read again.

Painkillers (analgesics)

Analgesics are some of the most widely-used drugs and there can be few people who have not taken an aspirin or two at some time in their lives. Though there are several ways of relieving pain, two important types of analgesic are the anti-inflammatory agents, such as aspirin, and the opiates which act by affecting pain perception in the brain.

The anti-inflammatory analgesics are discussed in relation to rheumatic diseases where they are particularly valuable. They are also the drugs of choice for the many pains caused by inflammation in any site, but are frequently effective when inflammation is not an obvious feature, as in the case of the common type of headache. Their chief adverse effect, inextricably bound up with their mode of action, is irritation of the stomach. However, many people have no trouble of this sort. For those with rheumatic or other disorders and who need large doses of and prolonged treatment with anti-inflammatory analgesics, there are many alternatives which may cause less gastric irritation or be more potent, as discussed in Chapter 11.

Paracetamol, probably has a similar mode of action to aspirin but appears mainly to block prostaglandin production within the brain. It does not therefore irritate the stomach and can be taken by those who have had bad reactions to aspirin. The disadvantages of paracetamol are that it is less effective in controlling inflammatory pain and in overdose can cause severe or fatal liver damage. Quite mild overdose (even as few as 15 tablets) can be dangerous in some persons, particularly as the effects may not become apparent until a few days later when it may be too late for effective treatment.

Opiates diminish the perception of pain but also affect mood to a variable degree. Major opiates such as morphine and heroin are particularly effective pain-killers and are also potent tranquillizers. This latter property can be valuable for patients with advanced cancer and also for those with the acute and agonizing pain of a heart attack. In the latter case, the resulting anxiety can greatly worsen the irregularities of the heartbeat; morphine or heroin are therefore the first line of treatment .

The other side of the coin is that the mood-elevating effect of opiates makes them addictive. Other opiates such as codeine or dihydrocodeine are much less potent pain-killers and also have little effect on mood. There is therefore little likelihood of dependence and these drugs are widely used, often in combination with other analgesics such as aspirin or paracetamol or both.

Common adverse effects of opiates (apart from dependence) are nausea and reduced bowel motility leading to constipation. These drugs are therefore frequently used for diarrhoea. They are also useful for suppressing cough.

Drug dependence (addiction)

Drug dependence is the inability to control craving for a drug and an overwhelming need to take it to maintain a sense of well-being. Further, stopping the drug causes severe mental and sometimes dangerous physical illness.

The most common cause of serious drug dependence in Britain is alcohol. Though moderate social drinking is virtually harmless, over-indulgence becomes socially disruptive and causes deterioration of performance at or loss of work, accidents, violence, and often break-up of family life. Sudden cessation of drinking can cause acute anxiety, tremor, and fits. Delirium tremens is a form of this withdrawal state characterized by terrifying hallucinations. The causes of alcoholism are unknown. Contributory factors are its availability and pleasurable tranquillizing action, but there may also be a genetic factor which makes some persons more susceptible.

Similarly the causes of dependence on other drugs are unclear, apart from the fact that drugs of dependence produce a pleasurable change in mood. Such drugs are of two types, namely those such as alcohol and opiates which are sedating (tranquillizing), and those such as cocaine and amphetamine which are stimulants and give rise to sensations of euphoria, excitement, and extroversion—the life-and-soul-of-the-party, as it were.

It is frequently asserted that addiction to drugs such as heroin is a form of escape from poor social conditions or unsatisfactory family life. The less charitable tend to ascribe it to inadequacies in the addict's personality and consequent failure to cope with everyday life. Whether or not any of these factors contribute in some cases is unclear and we know little about the fundamental causes.

In addition to the hazards of withdrawal, which can be fatal, the complications of drug abuse are many as morphine or heroin can cause coma and death from respiratory failure. Overdose of cocaine can cause sudden death from convulsions or over-stimulation of the heart.

Withdrawal of cocaine does not seem to lead to a physical withdrawal illness, but prolonged abuse of the drug can cause severe deterioration of personality. Injection of addictive drugs brings with it the hazards of infection, particularly hepatitis or AIDS, but also many others which may be as dangerous.

The indirect effects of drug abuse include widespread crime to obtain money for drugs as well as that committed by the drug peddlers to protect themselves or their profits. Violent crimes can also result from the use of hallucinogenic drugs, such as LSD, or more frequently from alcohol. Alcohol is also a major factor in road traffic accidents.

5

Disorders of the heart and circulation

Diseases of the heart and blood vessels are the most common causes of death in the West, yet some of these conditions are relatively rare in developing countries. The reasons for such differences are not fully understood, but there may be aspects of our way of life which increase the risk of developing heart disease. Changes in life-style could therefore bring significant benefits.

Heart attacks (coronary thromboses)

The heart pumps blood around the body to supply it with oxygen and nutrients, and to remove harmful waste products. It is a muscle which repeatedly contracts and relaxes in a rhythmical fashion to provide this pumping action. Because of this work which continues throughout life, the heart itself requires oxygenated blood and glucose. These reach the heart muscle via the coronary arteries. Blockage of these arteries is the usual basis of heart attacks and of angina pectoris. In a heart attack, blockage of a coronary artery may be so complete that the part of the heart muscle which normally depends on that vessel for its blood supply ceases to function and dies. The area may eventually be replaced by fibrous scar tissue, but does not regain the ability to contract and assist in the pumping action of the heart.

The most dangerous period during a heart attack is the first few hours. The pain is typically appallingly severe, the victim often has a sensation of impending doom and is deeply anxious. The pain and anxiety further strain the heart, and therefore need to be controlled with powerful pain-killing drugs. Such agents are the opioids (derivatives of opium), which include morphine and heroin. When used in emergencies such as heart attacks they carry no risk of addiction. An alternative is the less powerful nitrous oxide ('laughing gas') and this is frequently used by emergency ambulance crews when taking the patient to hospital.

A dangerous complication which can develop in the early hours of a heart attack is that the heartbeat becomes erratic. If the heart beats too fast or too slowly, or becomes irregular, then the circulation may be insufficient to support life. A regular rhythm can usually be restored if treatment is begun as soon as the abnormality starts.

Drugs used to restore normal heart rhythm are many and diverse in their modes of action. Examples are lignocaine, verapamil, amiodarone, and beta-blockers. The choice depends on analysing the nature of the disturbance of rhythm using an electrocardiogram, which records the quality of activity of each part of the heart. In severe cases, however, the disturbance of rhythm is so severe that is is lethal if not immediately stopped. This has to be done by applying a momentary electric current to the chest to shock the heart back into normal rhythm.

Prevention of heart attacks or recurrence

Any of us may be at risk from heart attacks—even a few apparently healthy athletes—but once a person has had a heart attack, the risk of another is even higher. Aspects of life-style which increase the risk of heart attack are smoking and, probably, some aspects of diet and lack of exercise. Diabetes may also predispose to heart disease. Another important factor contributing to heart attacks and other kinds of heart disease is high blood pressure (hypertension) which in turn is often related to obesity. Over-indulgence in alcohol may also contribute. Denial of many of the pleasures in life may be the price that has to be paid for lessening the risk of a heart attack.

Hunter–gatherer populations and many other people with comparable life-styles where there is hardly ever enough food do not suffer from this type of arterial disease. The reason for this difference is not known, but in Western man physical inactivity and overeating may be contributory factors.

Physical exercise not only helps to burn up excess food but also increases the rate at which a fatty material called cholesterol leaves the artery wall.

Deposits of cholesterol in the lining of the coronary arteries are an important cause of blockage (artherosclerosis or atheroma). In Western man such deposits are first seen in apparently healthy people from adolescence onwards and may progress in severity with age. This process is accelerated in smokers. Both dietary and genetic influences control blood cholesterol. You cannot select your parents, but your diet at least can be changed. Reducing animal and dairy fat but increasing fish and vegetable oil in the diet can lower blood cholesterol levels dramatically. In a minority of people, diet alone will not produce an adequate response, and although it may help to remain on a cholesterol-lowering diet drugs will be needed. Most of the blood cholesterol is manufactured in the liver, and drugs such as clofibrate (Atromid-S), bezafibrate (Bezalip), and nicotinic acid derivatives reduce the amount of cholesterol made by the liver. Cholestyramine (Questran) and similar resins reduce the absorption of cholesterol compounds by the intestine.

Another way of reducing the risk of a recurrence of heart attacks is by giving a drug which blocks the nervous stimulation of the heart. These sympathetic nerves have many actions, but in particular increase the excitability of the heart (beta effects). Drugs which selectively block the actions of these nerves on the heart are called beta-blockers. Examples are propranolol (Inderal) and metoprolol (Betaloc; Lopresor). If one of the beta-blockers is taken continuously after a heart attack, then the chance of another heart attack is appreciably reduced. Beta-blockers are also used to lower high blood pressure. Unfortunately these drugs also have disadvantages in that they make asthma worse and frequently cause lassitude, drowsiness, or even depression.

The cells in the blood which are involved in clotting are platelets. When these become stuck together, blood clots can form and block the arteries (occlusive vascular disease). Drugs which reduce the stickiness of the platelets can reduce the chance of blood clots forming in the coronary arteries and may also slow down the deposition of cholesterol in the walls of arteries. Such a drug is aspirin, which appears to reduce the chance of heart attacks if taken in low doses over a long period.

Angina pectoris

In this condition attacks of chest pain are brought on by physical effort or emotional excitement. The pain of angina is, however, short-lived, because as soon as the work of the heart is reduced (by stopping exercise or becoming calmer) then the blood supply to the heart is sufficient. The pain comes only if the blood supply is less than the needs of the heart. Angina is due to the same type of arterial blockage which is responsible for heart attacks, but to a less severe degree so that pain is only experienced when the needs of the heart are increased.

The management of angina includes removing causes of arterial occlusion. Thus cigarette smoking, lack of regular exercise, hypertension, and high blood cholesterol levels may be treatable aggravating factors.

A variety of drugs can stop the pain of angina. Nitrates such as glyceryl trinitrate can rapidly stop an attack or prevent one if taken before exercise. Glyceryl trinitrate is taken as a tablet which is crushed between the teeth and held in the mouth. The drug is absorbed directly through the lining of the mouth into the circulation and therefore starts to work rapidly. It acts by opening up blood vessels in the body and so lessens the work of the heart because the resistance to blood flow is lowered. The main toxic effect is that large doses can produce a throbbing headache.

Propranolol has been mentioned as a drug which blocks the stimulatory nerves to the heart and thereby reduces the risk of recurrence of a heart attack. The group of drugs to which propranolol belongs, the beta-blockers, is also valuable in the prevention of anginal attacks. These drugs reduce the need of the heart muscle for blood because the heart works less hard during exercise or emotional arousal. Patients who take this type of drug often notice that they find it difficult to exercise, and joggers may be slowed down. A much more serious toxic action, as mentioned earlier, is that asthmatic or bronchitic patients may become very breathless. Anyone with these respiratory complaints should therefore avoid taking propranolol or any similar drug such as oxprenolol, metoprolol, or atenolol.

Nifedipine and verapamil belong to a third group of drugs which prevent angina by relaxing arteries, including the coronary arteries.

These anti-anginal drugs can reduce or abolish anginal pain, but do not unblock coronary arteries obstructed by deposits of fatty materials. As a consequence, changes in life-style are important to slow down the progression of disease.

High blood pressure (hypertension)

Hypertension is an excessively high pressure of the blood in the arteries. This is usually due to constriction of the fine vessels through which the blood passes when leaving the arteries. Although this can be caused by disease in other parts of the body, particularly the kidneys, in over 80 per cent of patients the cause of the raised blood pressure is unclear. It seems to be a tendency which is inherited from both parents (essential hypertension).

Hypertension does not, in itself, cause symptoms, but if unusually severe may cause headaches. The main reason why it is necessary to lower raised blood pressure is to prevent the complications which can result. These include heart attacks, angina or strokes (haemorrhage or blood clot in the brain), and straining the heart muscle. High blood pressure also contributes to atherosclerosis and other arterial diseases.

There are many drugs which will lower a raised pressure. These include substances which cause the body to lose water and salt into the urine (diuretics), drugs which block the action of the nerves which normally stimulate the circulation, and drugs (vasodilators) which directly open up the small blood vessels which are excessively tight in this condition.

Diuretics

Diuretics increase the loss of water from the body by increasing urine production. They include bendrofluazide (Aprinox) and cyclopenthiazide (Navidrex). Less useful are frusemide (Lasix) and bumetanide (Burinex). They lower the blood pressure by a combination of mechanisms: reduction in the volume of blood that has to be pushed around the arteries, loss of salt, and a direct relaxing effect on blood vessel walls. The loss of salt (sodium chloride) contributes to the lowering of blood pressure, but excessive salt depletion can cause feelings of weakness, or even fainting. The elderly are particularly sensitive to salt loss.

Another chemical which may be lost in excessive amounts is potassium. In some patients this may have to be replaced by modifying the diet (increased amounts of citrous fruits and tomatoes) or by taking potassium tablets (Slow-K or Sando-K). Diabetes and gout can be worsened by such treatment.

As with many of the antihypertensive drugs, some diuretics, particularly the thiazides, may cause impotence. This is a subject which patients may not mention, but if during any type of drug treatment, a man's sexual performance deteriorates, he should report it to his doctor. Alternative drugs can usually be found.

Nerve blockers

Nerve blockers influence the nerves controlling the heart and blood vessels. Beta-blockers and their toxic effects have been mentioned earlier and include propranolol, metoprolol, and atenolol. These drugs usually lower a raised blood pressure, but the maximum benefit may not be reached until the drug has been taken for six or more weeks. Other nerve blockers include prazosin (Hypovase) and indoramin (Baratol).

Vasodilators

The vasodilators can sometimes cause headaches, but this group includes the most effective antihypertensives available such as the calcium antagonists. Examples are nifedipine (Adalat) and verapamil (Cordilox). They have the useful property of being more active in severe blood pressure than in mild hypertension.

The value of treatment of hypertension

The general results of lowering a raised blood pressure with drugs are very promising. The risks of heart attacks, strokes, and heart failure which accompany hypertension are all reduced if the pressure is kept down to normal levels. Severe hypertension should therefore be treated. However, the benefits of drug treatment of mild hypertension are less clear—all drugs have toxic effects and the risks due to a slightly raised blood pressure *may* be less than those of drugs. There is therefore a need for alternative methods of lowering the blood pressure. Losing weight is the first prerequisite for many. Physical exercise and relaxation techniques each produce a modest but significant lowering of blood pressure. The exercise has to be performed for at least 30 minutes three times weekly and be vigorous enough to make the heart beat faster. Jogging, brisk walking, or swimming are suitable forms of exercise. Exercise, however, should be started gently by anyone who is out of condition. Occasionally even apparently healthy young or youngish people (in their 40s) drop dead during jogging as a result of symptomless, unrecognized heart disease. Relaxation techniques (sometimes misnamed transcendental meditation) also have to be performed several times each week, and involve the expenditure of at least 30 minutes at each session. Tape recordings of instructions in relaxation are helpful, and cheaper than hiring an instructor.

Oedema (dropsy) and breathlessness

In many normal people and in some with diseases of the heart and circulation, fluid accumulates beneath the skin to produce swelling (oedema) which can be seen and felt. The fluid obeys the laws of gravity and so settles in the lower parts of the body. Thus, a person who is standing or walking for long periods, may develop swelling around the feet and ankles. When this happens in normal people, no treatment is necessary. If it is due to local conditions in the legs (such as varicose veins) or to prolonged sitting (as happens on long aircraft journeys or in the chairbound elderly) then the cause can be attended to. However, if heart disease is the cause, then the accumulation of fluid may become so severe and widespread as to involve the lungs. Because the lungs are organs of absorption of oxygen, increased fluid in the lungs causes breathlessness. In heart disease, increasing breathlessness on effort may be an early feature—initially noticeable with more strenuous activities, such as climbing upstairs, but as the condition worsens, severe breathlessness can be experienced even at rest. Under these conditions, diuretic drugs can help to remove excess fluid. Not only will the swelling of the ankles be reduced, but breathlessness should be improved.

Palpitations

Awareness of the sensation of the beating of the heart is called palpitations. It is frequently experienced by normal people during exercise or heightened emotion (such as anxiety or anger) and is due to the heart beating unusually powerfully or quickly. Alternatively, it can be due to a person who is particularly sensitive and introspective, becoming anxious. Sometimes palpitations can be a symptom of an abnormal rhythm or rate of the heartbeat. Most palpitations are experienced by people with normal hearts and require no treatment. Many drugs (including propranolol, digoxin, verapamil, disopyramide, and mexiletine) are available for the treatment of abnormally rapid rhythms of the heart. However, the choice of these drugs is usually a matter for a specialist cardiologist as drugs of this type are amongst the most dangerous used in medicine.

Poor circulation

The function of the heart is to pump blood throughout the body. If there is narrowing of the vessels, blood may not reach the more remote parts of the circulation such as the fingers and toes and parts of the limbs, even though the heart continues to pump normally.

The form of poor circulation in the hands and feet which is much less sinister than the above, consists of attacks of pallor and blueness of the fingers and toes following exposure to cold. This is called Raynaud's phenomenon, and is common in women. It is due to temporary spasm of the fine arteries in the digits. Drugs are usually unhelpful in preventing this, and it is more effective to keep the whole body warm and to avoid as far as possible putting the hands (or feet) into cold water. Woollen gloves and socks should be worn in cold weather.

The narrowing of the vessels is also frequently due to atherosclerosis, the common form of degenerative arterial disease. If this happens in the major arteries of the leg, it can cause pain in the calf muscles during exercise. The mechanism is similar to that in angina pectoris, in which narrowing of the coronary arteries causes chest pain on exercise. In both conditions, cigarette smoking aggravates the disease because nicotine accelerates the formation of atheroma and also causes spasm of the arteries. In both diseases high blood cholesterol contributes and therefore cholesterol-lowering diets and drugs may be helpful.

6

Disorders of the digestive system

The gastro-intestinal tract includes the mouth and teeth, the throat (pharynx), the oesophagus which connects with the stomach, the small intestine where most food and drugs are absorbed and the large intestine which is mainly concerned with dealing with the residues which leave the body via the rectum and anus.

The stomach secretes large amounts of hydrochloric acid, from which its lining is protected by complex mucous substances. This acid is essential to health as it destroys many swallowed microbes and also starts off the process of digestion with the help of enzymes which break the food down into absorbable molecules. Failure of production of gastric acid is the result of loss of specialized cells of the stomach lining and this can lead to several diseases. It may, for example, increase susceptibility to stomach cancer. Secretion of acid and the digestive enzyme, pepsin, is partly controlled by the nervous system via the vagus nerve. It is not surprising therefore that the activity of the stomach is strongly influenced by emotion: anger has been shown to cause the stomach lining to become engorged and inflamed. The term 'nervous stomach' is not without justification therefore as some people are very prone to indigestion whilst others are happy to subsist on fish and chips washed down with unlimited amounts of beer.

Another important organ of the digestive system is the pancreas, which produces enzymes and insulin which controls blood sugar. Bile contains enzymes which are particularly important for the digestion of fats. Finally, the digestion products of food pass to the liver where they are processed into a great variety of substances related to virtually all bodily activities.

Mouth and throat infections

Care of the teeth has been dealt with fully in another Oxford paperback (*Good Mouthkeeping* by John Besford).

Dental abscesses

A dental abscess can form when decay has spread into the pulp (nerve) of the tooth and down its root. In most cases this is chronic and painless but it can be acute and painful. It is possible to drain the infection from the tooth and carefully fill the root with inert material which will block the re-entry of bacteria. No other treatment of the root abscess may then be necessary, and the tooth, though 'dead', can remain useful for many years. Antibiotics are rarely needed.

Similarly antibiotics are not normally needed if there is an acute abscess and the tooth has to be extracted. The extraction drains the infection away far more quickly than any antibiotic can act and any possible side-effects of the antibiotic are avoided. If, however, the tooth cannot be extracted straight away it is common practice to give an antibiotic such as penicillin or metronidazole to keep the infection under control and perhaps lessen the pain. This helps only if the causative bacteria are sensitive to these drugs but an analgesic such as aspirin or paracetamol in adequate doses is usually also necessary.

Systemic effects of dental infections

Painless, low-grade infections round the teeth were in the past thought to cause a variety of diseases in distant parts of the body, ranging from arthritis to disorders of sight. This idea, termed 'focal sepsis', led to the extraction of innumerable teeth in many patients without any benefit. It is now obsolete. However, enormous numbers of bacteria are harboured in pockets round the teeth and these bacteria are released into the bloodstream by many dental procedures, toothbrushing, or chewing food. This is normally harmless, but if great numbers of these bacteria are released, particularly by extractions, they can settle on heart valves which have been damaged by previous disease or which have congenital defects. The result can be a rare heart infection known as infective endocarditis and can lead to death. For people who have heart defects therefore it is desirable to give antibiotics such as amoxycillin before certain dental procedures, such as extractions or operations on the gums, as a preventive measure. However, there are many other possible precipitating causes of this disease (dental procedures account for only a small minority of cases) and on many occasions the disease has no apparent triggering factor.

Oral mucosal infections

Infections of the mucous membrane lining the interior of the mouth are uncommon. The main kinds are infection with *Candida albicans*, the fungus which causes thrush, and herpetic (viral) infections.

Thrush is a well-known infection in babies. It can cause severe infections in debilitated patients and was aptly called by a nineteenth century physician, a 'disease of the diseased'. Oral thrush is an early characteristic of AIDS in adults or during treatment with immunosuppressive drugs, as consequence of loss of normal immune defenses.

A traditional treatment is with gentian violet, but this can ruin clothes and may even cause a chemical burn in the mouth. More modern alternatives are nystatin (as a liquid suspension or pastilles) and amphotericin (as lozenges or a syrup). These drugs are specific antifungal antibiotics and though they may not taste very pleasant have no significant side-effects as they only act locally and are not absorbed. Newer antifungal drugs are miconazole which can be used topically in the mouth in gel form and ketoconazole tablets. Unlike the older antifungals, these are absorbed from the gut which makes them more effective but can have side-effects. Very occasionally, ketoconazole has caused liver damage after long courses of treatment for more intractable fungal infections.

Herpes simplex virus is a notorious cause of sexually transmitted vaginal infections but another type of this virus (type I) can cause painful mouth infections. This used to be common in infancy and was often thought to be 'teething'. It is now just as likely to affect adults. Oral herpes causes minute blisters which burst, leaving painful ulcers; the glands in the neck become swollen and a febrile illness often develops. There are few useful antiviral drugs, but acyclovir (Zovirax) is an exception and remarkably effective for herpetic infections if given early enough. Tablets and a liquid preparation are available.

Cold sores (fever blisters) are recurrent herpes infections which affect the borders of the lips. Those who suffer from them will recognize the tingling or burning sensations which herald an attack. A succession of blisters then appears and these are followed by unsightly crusting. The attack can be greatly shortened by applying acyclovir cream to the area as soon as the warning symptoms are felt. If treatment is delayed until blisters appear acyclovir will not repair the damage that the virus has already done but may cause fewer new blisters to form. Acyclovir is a remarkably safe drug but the ointment may cause brief local stinging. Cold sores are, incidentally, infectious and can cause herpetic infection in someone else by kissing for example.

Another herpes infection is shingles (herpes zoster). This usually affects the chest, but can involve the face, mouth, and eyes. Facial shingles causes similar effects to those of herpes simplex, but involves the skin of one side of the face as well as the mouth, and pain which may be felt as toothache. If the forehead is affected then the eye is also involved and this can affect sight. Acyclovir is effective in shortening the illness but must be given as soon as symptoms start and in large doses as shingles is a debilitating illness, especially in the elderly. Side-effects are usually minimal but the dose should be reduced in those with failing kidneys. In very ill patients, such as those having organ transplants, acyclovir has to be given intravenously, and is frequently life-saving.

Sore tongue

Sore tongue is a common complaint even when the tongue appears healthy. This can be the earliest symptom of anaemia, which is readily confirmed by a simple blood test. Vitamin deficiencies are not a significant factor in the West and frequently no explanation can be found. In some cases it may be a symptom of some internal stress. Antidepressive treatment sometimes provides relief.

Sore throat

Sore throat (pharyngitis) (See Chapter 3), is usually caused by any of a great variety of viruses for which there is no specifically effective treatment. Symptoms may be lessened by use of gargles (for example, aspirin dissolved in warm water), but the infection usually has to be allowed to take its course.

Indigestion (dyspepsia)

This catch-all term covers abdominal symptoms such as fullness of the stomach (bloating), wind, vague discomfort or pain under the diaphragm, and heartburn. Indigestion is not usually due to a definable disease such as a gastric ulcer (though peptic ulcers can cause similar symptoms), and X-rays show no abnormality.

Causes of indigestion include hurried meals, ill-chosen food (particularly too much fried food—continued re-use of frying fat produces highly irritant acids), and 'stress' (anxiety)—the word 'dyspeptic' is (or was) used for an impatient, irritable temperament. The scale of the problem is vast: NHS prescriptions for antacids recently cost over £12 million in Britain in one year alone; possibly as much again is spent on over-the-counter remedies.

Most people learn to avoid particular foods or any other cause of the trouble. However, they may also have to resort to antacids, the most common type of indigestion remedy. These drugs probably mainly work by speeding up emptying of the stomach rather than by merely neutralizing gastric acid. Sodium bicarbonate (a traditional remedy) acts rapidly but has the disadvantage that it is likely to produce more wind with consequent belching. It has serious adverse effects if too much is taken. Currently, the most satisfactory antacids are combined magnesium and aluminium compounds. These mixtures are long-acting and have the advantage that magnesium hydroxide alone is prone to cause diarrhoea, while aluminium hydroxide has the opposite effect. A great variety of other substances may also be added. These include dimethicone to lessen flatulence, alginates to protect the lining of the stomach and oesophagus, and many others but it is difficult to confirm that they confer any additional benefits.

Antacids should be taken when symptoms are expected—after meals or at bedtime for example—and act most rapidly when taken in liquid form, though tablets are more convenient for daytime use. They should not be taken with other drugs as they frequently interfere with absorption. They should only be taken in sufficient quantities to control the symptoms.

Part of the pain of indigestion may be due to spasm of the muscles of the digestive tract, and antispasmodic drugs such as belladonna alkaloids or newer agents such as ambutonium or poldine are included in several preparations for dyspepsia.

In a minority, indigestion is a symptom of serious disease such as peptic ulcer or stomach cancer. For those with intractable indigestion or those in whom indigestion starts unexpectedly, it is important to have X-rays and direct examination of the interior of the stomach by endoscopy.

Oesophageal reflux and hiatus hernia

Reflux of gastric acid (heartburn) is part of the various symptoms conveniently called 'indigestion'. It can be particularly troublesome when part of the stomach is drawn up beside the oesophagus. This is known as hiatus hernia. It can be detected by an X-ray and may cause severe heartburn, especially when bending forwards. Nevertheless, many patients found to have hiatus hernias have no symptoms and many that have acid reflux have no hiatus hernia.

For those with acid reflux, the main measures are first, to lose weight, second to use whichever indigestion remedies are found to be most effective, third to avoid fatty or acid foods, or alcohol in the evening, and fourth to tip the bed with blocks under the head end.

Gastritis

Acute gastritis (inflammation of the stomach lining) results mainly from ingestion of irritants (particularly alcohol and aspirin) or infections. Symptoms range from loss of appetite (anorexia) and a sense of bloating to nausea and vomiting or even bleeding from the stomach.

It most cases it is sufficient simply to take as much fluid as necessary and avoid food for a day or so. Viral infections affecting the gastro-intestinal tract have to be allowed to take their course as there is no specific treatment. Irritant chemicals swallowed accidentally or deliberately need expert care and this may include immediate wash-out of the stomach in hospital.

Peptic ulcer

Peptic ulcers form in the stomach or duodenum but, contrary to a widespread belief, are not usually the result of 'too much acid'. They are, instead, a failure of the mechanisms protecting the stomach and duodenal linings from the normally produced acid. In many cases there is no readily recognizable cause, but genetic factors, smoking, 'stress', aspirin or related pain-killers, and corticosteroids can be contributory.

Peptic ulcers frequently cause indigestion; alternatively, there may be complete absence of symptoms until there is a sudden complication such as a severe haemorrhage with vomiting of blood or a perforation. Stomach ulcers typically cause pain at varying times after meals. Traces or larger amounts of blood are passed in the stools and the ulcer can often be seen on an X-ray. However, the diagnosis is best confirmed by directly visualizing the ulcer after passing an endoscope into the stomach. Duodenal ulcers, by contrast, typically cause pain several hours after meals, if a meal is missed, or at night. This pain is relieved by food or antacids.

The first treatment of peptic ulcer is to keep clear of known aggravating factors. Stopping smoking and bed-rest are virtually the only measures, other than drugs, that hasten healing of ulcers.

Drug treatment for peptic ulcers is usually unavoidable. Since the stomach cannot defend itself against its own acid, the general principle is to neutralize or block acid production. Antacids frequently relieve symptoms but do not induce healing of ulcers unless taken frequently, in large amounts. To do so may be inconvenient and may in the long-term cause toxic effects.

The most rapidly effective ulcer-healing drugs are a type of antihistamine, different from those used for allergies. These block the histamine targets (H2 receptors) that mediate acid production in the stomach wall. Cimetidine and ranitidine are the main examples of antihistamines which effectively block gastric acid production. These drugs usually cause few significant side-effects, but treatment (one or two tablets at night) may have to be maintained to prevent relapse. They are frequently also effective for hiatus hernia or persistent acid reflux.

Possible adverse effects of cimetidine, though uncommon, include antagonism of male sex hormones and breast growth in men (gynecomastia). It may also cause confusion in elderly patients and interfere with the metabolism of several other drugs. Ranitidine lacks most of these side-effects apart from possibly causing confusion in the elderly.

Another type of ulcer-healing drug is carbenoxolone. This drug is chemically related to a derivative of liquorice (a traditional remedy for peptic ulcer) which probably acts by improving the mucous barrier protecting the stomach lining. Though moderately effective, carbenoxolone causes troublesome side-effects, notably retention of fluid and sodium but loss of potassium. These can cause or worsen heart failure, hypertension, oedema (with concomitant increase in weight), and irregularities of the heart. This drug is therefore unsuitable for older patients and is less useful than the H2 antihistamines. Sucralfate (Antepsin) and bismuth chelate are newer anti-ulcer drugs which probably also help to improve protection of the stomach lining. They appear to be very safe and effective. Pirenzepine (Gastrozepin) acts by inhibiting nervous stimulation of acid secretion. It is as effective as cimetidine or ranitidine and can be used in conjunction with either of them in resistant cases. Pirenzepine may cause some drying of the mouth or mild visual disturbances.

Nausea and vomiting

These common complaints can result from disorders of the gastrointestinal tract itself or from influences external to it; there is therefore a great variety of causes and no single reliable remedy.

Infections

Many viral infections (such as 'gastric flu') can cause vomiting and/ or diarrhoea as part of a generalized illness. There is no specific treatment, but a good fluid intake should be maintained.

Food poisoning in medical terms is infective gastro-enteritis and results from eating food contaminated by microbes or their toxins. Despite attempts at improving catering hygiene, food poisoning is very common, particularly where food is kept warm for long periods. If the food has been contaminated with bacteria by a food handler, for example, this allows the bacteria to multiply sufficiently to cause trouble. Salmonella are among the most common bacterial causes and not infrequently infect chicken carcasses.

The diarrhoea suffered by travellers is due to infections such as these or to the toxins from various bacteria. The frequency and severity of these infections is roughly proportional to the warmth of the climate but is more directly dependent on the local standards of hygiene. The Far East and Mexico are particularly high-risk areas.

Antibiotic treatment is neither particularly effective nor advisable because the antibiotics do not act quickly enough and in any case may not destroy the particular microbes causing the trouble. In addition, many antibiotics can cause nausea or even vomiting and overuse of these drugs promotes the proliferation of antibiotic-resistant bacteria. In Mexico such resistant bacteria have caused huge epidemics with many deaths.

Most cases of food poisoning should therefore usually be allowed to take their normal course, but plenty of fluids should be taken. In mild cases of food-poisoning, diarrhoea remedies can be convenient for controlling symptoms but should be avoided for the more severe cases as they will bottle up the infections within the gut.

Motion sickness

Motion sickness is due to disturbance of the semicircular canals of the inner ear. As with gastro-intestinal infections, there is wide individual susceptibility and a strong psychic component. The latter is suggested by the fact that while passengers in a vehicle may be affected, the driver who one hopes is concentrating on his job often is not. Also, in the worst cases, nausea and vomiting can start even before the journey begins and persist afterwards. Motion sickness can usually be controlled by taking advantage of a side-effect of conventional antihistamines such as dimenhydrinate (Dramamine) many of which are available without prescription. Hyoscine is possibly even more effective, but both this and the antihistamines cause a dry mouth and drowsiness. This sedating effect possibly enhances their anti-emetic action on the vomiting centre in the brain. These drugs should be taken 30 to 60 minutes before travel but car drivers should not take them.

Pregnancy

Though vomiting in pregnancy is a very troublesome complaint, drugs should be avoided because of the possibility of affecting the fetus. The first measures are therefore care with the diet and reassurance. There is, however, minimal evidence that anti-emetic drugs are harmful to the fetus and if nausea and vomiting are severe, antihistamines such as meclozine (Ancoloxin) or promethazine (Avomine) may be given by some obstetricians.

Drug-induced nausea

Many drugs can cause nausea and/or vomiting either because they are gastric irritants or because they stimulate the vomiting centre in the brain; some act by several such mechanisms.

Anticancer drugs, particularly cisplatin, can cause distressingly severe nausea even when given by injection. Several drugs are now available to overcome such problems.

The phenothiazines are chemically related to the antihistamines. Their main use has been in the control of schizophrenia (Chapter 4), essentially because of their sedating action, but they also block the vomiting centre in the brain. Examples are prochlorperazine (Stemetil; Vertigon) and chlorpromazine (Largactil) which are used to control nausea caused by drugs, radiation, and anaesthetics.

Post-operative vomiting is frequently due to the use of morphine for premedication; many anaesthetists claim that post-anaesthetic nausea can be avoided by using benzodiazepines (Valium-like drugs) for premedication.

Metoclopramide (Maxolon) is one of the most effective anti-emetics, in that it both promotes normal forward movement of the stomach contents and increases the tone of the sphincter (valve) between the oesophagus and stomach as well as blockading the vomiting centre in the brain. This central action is, however, essentially the same as that of the phenothiazines and a distressing effect is a tendency to provoke involuntary muscle activity such as spasm of the jaw or limb muscles, particularly in the young. Prolonged treatment should not be given because permanent disorders of movement can result.

Domperidone (Motilium) also acts on the vomiting centre and, because it is less prone to cause sedation or involuntary movements than metoclopramide or the phenothiazines, is a useful anti-emetic.

Haloperidol (Serenace), another type of drug used for treating psychoses has phenothiazine-like activity in controlling vomiting but is less sedating and longer acting.

Despite this wide choice of drugs, there is sometimes no simple answer to the distressing nausea associated with anticancer treatment and a combination of anti-emetics which usually includes domperidone and/or metoclopramide, prochlorperazine and haloperidol has to be tried until the patient is comfortable. A benzodiazepine such as lorazepam may also be included and among other benefits the additional sedation this provides is helpful.

Ménière's disease

This is a disorder of the semicircular canals and causes hearing loss with tinnitus (ringing in the ears), vertigo ('dizziness') and nausea comparable to that of motion sickness. Antihistamines are helpful for nausea from this cause but betahistine (Serc) is better as it also relieves the vertigo. Betahistine is virtually only useful for Ménière's disease as it is (perversely) also capable of causing nausea and can aggravate asthma (by histamine-like activity) and peptic ulcer.

Intestinal diseases

The most common complaints caused by these diseases are diarrhoea, constipation, or abdominal pain. Less common are diseases of the small intestine which typically cause impaired absorption and various forms of malnutrition. Coeliac disease, a condition caused by a reaction to gluten, is an example.

Crohn's disease

This is a relatively rare disease which mainly affects the lower end of the small intestine but can extend to the anus or involve the mouth. The cause is unknown and the symptoms are very variable but intermittent diarrhoea and constipation, abdominal pain, and fever are common. Later, anaemia or other deficiencies can result from impaired food absorption and in severe cases there can be uncontrollable painful diarrhoea mimicking ulcerative colitis. Rashes, abscesses and joint pains can be associated. Infection plays some part in the process and less severe cases can be maintained on sulphasalazine which is broken down in the bowel into a sulphonamide (antibacterial) and a salicylate which is anti-inflammatory. Other antibacterial drugs such as metronidazole may also be of some benefit but acute exacerbations and the more severe cases are only likely to respond to corticosteroids such as prednisolone. Troublesome joint pain is also only likely to respond to such treatment but once the acute phase has subsided the dose can be greatly reduced.

Supportive treatment and, in particular, a high-protein, high-fibre diet and treatment of any vitamin deficiencies are also necessary. It has to be admitted there is no cure and resection of part of a severely affected bowel may be necessary if all else fails.

Ulcerative colitis

This distressing disease which typically starts in early adult life is an inflammatory disease of the lower bowel (colon) which becomes extensively ulcerated. The cause is unknown but diarrhoea is the main effect; it is often deceptively painless but in severe cases can be painful with passage of blood and mucus in fluid stools. Anaemia, mild fever, and loss of weight are common.

Joint pains (arthralgia) and a great variety of other complications may be associated.Since the cause of ulcerative colitis is unknown, it has for long been believed that the disease has a strong psychic component. However, it would be a hard task for anyone with persistent, painful, debilitating, and uncontrollable diarrhoea to remain serene and full of *joie de vivre*.

Mild cases can be managed, like Crohn's disease with a nourishing high-residue diet, antidiarrhoeal drugs (see below), and sulphasalazine. Anaemia secondary to blood loss may need to be treated by giving iron. More severe cases require in addition prednisolone (by mouth or in an enema), whilst those with worsening disease need to be admitted to hospital for intravenous fluid replacement, antibiotics, and blood transfusion if necessary. Surgery is needed if the thinned bowel wall becomes grossly dilated (toxic megacolon) as it can rupture. Resection of the large bowel is also indicated in those with extensive long-standing disease as they have an increased risk of bowel cancer. This operation is curative.

Irritable bowel syndrome

This is the term given to symptoms which include abnormal bowel function (diarrhoea or constipation and mucus in the stools), abdominal pain, or dyspeptic symptoms. Anxiety or depression tend to be associated and symptoms are worse at times of stress and precipitated by various foods. Treatment has to be tailored to the individual patient. Avoidance of milk or milk products helps some and a high-fibre diet, with supplementary bran, is beneficial. Drugs which lessen intestinal spasm such as mebeverine (Colofac; Colven), dicyclomine (Kolanticon; Kolantyl), or propantheline may be useful. Their side-effects include drying of the mouth with consequent difficulty in swallowing, reversible disturbance of vision, and sometimes palpitations. This is because these drugs act through widespread nervous system (parasympathetic) pathways and do not have a sufficiently selective effect on the bowel.

Nevertheless the main treatment measures are to avoid troublesome foods, alcohol and smoking, and to limit tea and coffee consumption. Regular modest meals, exercise, and adequate sleep are highly advisable. Nevertheless, it may have to be accepted that just as some individuals are prone to indigestion and 'stomach troubles', others have over-reactive bowels which may be difficult to pacify.

Diverticular disease

The wall of the lower bowel can form bunches of small outpouchings (diverticulae) between the bands of longitudinal muscle, particularly in later adult life. Frequently this causes no symptoms but accumulation of faecal material can give rise to increasing inflammation (diverticulitis) and this can go on to inflammation of the surrounding tissues and abscess formation. The mass thus formed, together with abdominal pain and constipation (or less often diarrhoea), can mimic bowel cancer. In many cases, however, bowel diverticulae cause no symptoms and are no more than chance findings on X-rays.

Many believe that diverticular disease is another consequence of lack of fibre in the diet and uncomplicated cases require no more than a high-fibre diet and possibly, lubricant laxatives to retain healthy bowel activity. Complicated disease, involving attacks of acute diverticulitis require antibiotic treatment with such drugs as ampicillin or cefuroxime which act against many of the bowel bacteria. Ultimately surgery may be needed for perforations or long-standing masses with abscess formation.

Antibiotic-associated colitis

Many antibiotics can cause inflammation of the bowel as a result of destroying its normal bacterial flora and thus allowing other more troublesome bacteria to flourish. One otherwise safe and useful antibiotic, clindamycin is more prone to cause a severe form of this complication known as pseudomembranous colitis, with sloughing of part of the bowel wall. This has resulted in occasional fatalities but it must be emphasized that many other widely used antibiotics such as ampicillin can rarely cause this complication. Pseudomembranous colitis is caused by the bacterium *Clostridum difficile*. It responds to treatment with the antibiotics vancomycin or metronidazole given by mouth. Before the cause of this disease was known it seems that some of the deaths resulted from the giving of antidiarrhoeal drugs which bottled up the infection in the gut.

Dysentery: infectious colitis

Dysentery and other causes of infectious diarrhoea have, in the past, decimated armies and civilian populations alike. These intestinal infections can be caused by a wide range of bacteria and parasites, and are food- or (where sanitation is poor) water-borne. They no longer cause vast epidemics in developed countries because, even before their bacterial nature was understood, sewage systems (which keep excreta out of the drinking water) were built.

In most cases, the main treatment measure is to maintain fluid intake. In some of these infections, antibiotics such as tetracycline may shorten the illness but in others they can lengthen the period of infectivity and make the patient a potential disease carrier.

The common minor diarrhoeas of holidays abroad are also usually infective and a form of food poisoning. Popular remedies were traditionally based on morphine or codeine which among their many actions, reduce bowel activity and in the long-term cause severe constipation. These weak mixtures of morphine usually also contain kaolin (china clay) or calcium carbonate (chalk) to make the bowel contents less fluid. Although such mixtures are still widely used, more modern drugs for reducing bowel activity are loperamide (Imodium) and diphenoxylate (Lomotil). All of these, except loperamide, can cause undesirably deep sedation in children and in the elderly can cause faecal impaction and other complications. These antidiarrhoeal mixtures also do nothing for the underlying infection and might bottle-up the infection with serious effects, such as toxic megacolon as in ulcerative colitis. The possible hazards of such treatment have therefore to be balanced against the inconvenience of the diarrhoea. Many of these infections are so minor that antidiarrhoeal mixtures are useful to calm down the gut after the infection has taken its brief course.

In AIDS, chronic diarrhoea and wasting is so common a feature that in many parts of Africa, it is known as 'slim disease'. In many cases this diarrhoea is caused by protozoa for which there is no treatment.

Constipation

The Victorian obsession with bowel function has to some extent diminished. It has gradually become more widely appreciated that constipation does not lead to poisonous substances being absorbed from the bowel, and that habitual use of purgatives did more harm than good. Normal bowel function depends on fibre in the diet (to give adequate bulk) and enough physical exercise. In children and the elderly in particular, laxatives should be avoided. Constipation can be prevented by including bran-containing cereals, wholemeal bread and pasta, pulses, and fruit in the diet. Nevertheless, purgatives are widely used and are available in great variety.

Bulk-forming agents

These preparations make up for lack of bulk in the diet and are useful both for this purpose and for making the stools less fluid. They are therefore valuable in those with ileostomies or colostomies, diverticular disease, ulcerative colitis, and piles. In practice, these preparations are effective in certain types of diarrhoea as well as in constipation.

Examples of bulk-forming agents are bran (conveniently taken in foods such as biscuits or cereals), methylcellulose (which also acts as a faecal softener), ispaghula husk, and sterculia. Plenty of fluid should be taken with these preparations or, in extreme cases, the fibre can form a sufficiently solid mass to obstruct the gut. Overall, they are the most satisfactory laxatives, especially in the form of high-fibre foods, but their maximum effect may not be obtained until after a day or two.

Osmotic laxatives

These stimulate the bowel to act by absorbing water and so increasing the volume of bowel contents. A traditional example is magnesium sulphate (Epsom salts). Magnesium oxide, sodium sulphate, and lactulose (a non-absorbable sugar) act in the same way. These compounds act rapidly (2– 4 hours) and are suitable for occasional use.

Lubricants and softeners

Liquid paraffin is the traditional example but is no longer recommended: a newer drug is docusate sodium. These drugs are probably mainly useful for constipation associated with piles.

Stimulant laxatives

These include the traditional remedies such as castor oil, senna, and cascara, and also newer drugs such as bisacodyl. Overuse of them leads to weakening of bowel function and as a result, constipation. They are chiefly of use for medical purposes such as evacuation of the bowel before surgery, X-rays, or internal examinations (endoscopy). Their action is much more rapid than that of the bulking agents.

Liver diseases

The liver, among other functions, is the organ where most drugs are inactivated. Many drugs must be given in reduced doses or avoided altogether when there is serious liver disease. A few drugs can also cause liver damage, but this is far less common than say alcoholic liver disease.

Liverishness is no more than feeling off-colour—often as result of a hangover. The most obvious sign of disease involving the liver is jaundice, but frequently this is no more than a sign of obstruction of the normal outflow of bile into the intestine, by gallstones for example. In other cases, notably hepatitis, obstruction is secondary to liver disease. Most of the serious liver diseases are preventable rather than treatable.

Viral hepatitis

The most common type is hepatitis A, a food-borne infection that can readily be acquired on holiday in hot countries. Short-term protection can be provided by giving hepatitis A immunoglobulin. Though an attack can be unpleasant while it lasts, hepatitis A is unlikely to have any serious long-term consequences.

Hepatitis B is a much more serious disease. It is chiefly a risk to doctors and dentists who may be exposed to infected blood via accidental needle puncture for example. Hepatitis B should no longer be a risk from transfusion as blood is screened to exclude the virus. Intravenous drug addicts are, however, very much at risk, as are male homosexuals who often have an exceptionally severe form of the disease. The infection can also be spread heterosexually, though this is relatively rare except among clients of prostitutes, particularly in tropical countries. Otherwise the main risk to the general public if proper sterilizing procedures are not followed is from acupuncturists, tattooists, and the piercing of ears or other appendages.

Those at high risk, particularly health care workers, should be protected by active immunization with H-B-Vax which normally gives long-lasting protection and rarely leads to significant complications. As with other active immunizations, however, protection is not immediate and those who are likely to have been exposed to the virus should have hepatitis B immunoglobulin as well as H-B-Vax.

The severity of hepatitis B infection ranges from asymptomatic cases to fulminating disease quickly culminating in liver failure and death, but this is rare. The other serious complication is persistence of the disease process as chronic active hepatitis which can lead to cirrhosis. There is no antiviral drug specifically effective against hepatitis B, but treatment with alpha-interferon was recently found promising. This is active against many (in theory, all) viruses affecting humans but has to be given over a long period (several months) and then causes unpleasant toxic effects such as a persistent flu-like illness and inability to sleep.

Cirrhosis

Cirrhosis is the progressive destruction of liver tissue associated with inadequate attempts at regeneration and is a well-known consequence of chronic alcoholism. It can also result from hepatitis B and hepatoxic (liver toxic) drugs or chemicals. In many cases, however, no cause can be identified. Since there is no effective treatment, avoidance of risk factors is important as cirrhosis progresses inexorably to death from liver failure or liver cancer.

Gallstones and cholecystitis

Cholecystitis is inflammation of the gall-bladder where bile is stored. In the majority of cases, inflammation follows obstruction to the outflow of bile by gallstones. Stones are formed from cholesterol (an important component of the bile) and may also be calcified.

There are no known predisposing causes of cholecystitis apart from taking contraceptive pills, which raise the concentration of cholesterol in the gall-bladder.

Acute attacks can be precipitated by a fatty meal and cause abdominal pain, vomiting, and (sometimes) jaundice. Frequently, attacks subside if solid food is avoided (to lessen stimulation of bile secretion) and antibiotics are given. Pain is controlled with analgesics; but not opiates many of which increase spasm of the gall-bladder outlet.

Severe attacks with fever, jaundice, and persistent pain are best treated by surgery to remove the gall-bladder and impacted stones. Otherwise the gall-bladder may burst and cause more serious complications. As an alternative to surgery, for some of those who have recurrent attacks but the gall-bladder function remains good, cholesterol (but not calcified) stones may be dissolved by giving the drugs chenodeoxycholic acid or ursodeoxycholic acid. The former frequently causes diarrhoea and either can cause pruritus (itching). After disappearance of the gallstones has been confirmed by means of X-rays, long-term prophylaxis is needed as new ones can form after treatment is stopped.

7

Disorders of nutrition and the blood

Food affects our health at least as much as medicines. The main difference is that ill-advised eating habits are generally slower to show their effects. World-wide, however, malnutrition is the chief cause, direct or indirect, of ill-health and premature death. In the Western world, by contrast, overeating (not to mention drinking too much) is the main cause of obesity and it complications. Since it is depressingly difficult for those who put on weight too easily to maintain a necessarily spartan reducing diet, there is inevitably a demand for medicines which will help with this problem. Some foods, for example vitamins, can in certain circumstances be considered as medicines.

The nutrients essential for normal blood formation and function are obtained from the diet. Disorders of the blood are therefore discussed here, but it must be emphasized that many blood diseases, such as leukaemia, have no relation to diet.

Diet and health

Healthy eating depends on getting enough food (as fuel) to keep the body machinery going—but not too much. In addition, other substances such as iron and vitamins, are essential only in relatively minute amounts. But it must be emphasized that these do not normally need to be taken as extras— enough comes in a normal diet.

Despite strong opinions about it, the nature of a 'healthy diet' is controversial. It is virtually impossible to determine the effects of the diet in isolation and, even if it were, some of its effects may take a lifetime to become apparent. Diet has become an increasingly emotive issue and has sometimes led to food-fads.

Are Africans, for example, less prone to bowel cancer and other bowel diseases because of a high-fibre diet? Are Eskimos less prone to heart attacks because they consume so much fish oil? Do people of the Mediterranean coastal countries live longer because they consume more olive oil than other fats? These are only a few of the current questions, but in addition there are the ethical questions raised by vegetarians. Quite apart from the humanitarian aspects, there are certainly valid arguments that vegetarianism provides a more healthy diet. Meat is not essential for health: theoretically, it is an essential source of vitamin B_{12}, but vitamin B_{12} deficiency anaemia is surprisingly rare even among total vegetarians (vegans).

Fat

Whatever the theories of nutritionists and others, most of us reach the currently normal expectation of life on a mixed diet of meat, vegetables, and fruit. Obesity is nevertheless, a major health problem as a contributor to heart disease and premature death. Related to this is the level of fat-like substances, particularly cholesterol, in the blood. Major contributors to obesity and high blood cholesterol levels are excessive consumption of animal fats and lack of exercise. Animal fats include butter and cream but even lean meat has a significant fat content. These fats are, in chemical terms, 'saturated' fats. In general, hard fats such as mutton fat are highly saturated. Some vegetable oils such as sunflower or corn (maize) oil are 'polyunsaturates' and, in place of butter for example, may lower blood cholesterol levels to some degree. Olive oil (a monosaturate and previously not thought to be beneficial) appears now to be even more effective than polyunsaturates in lowering blood cholesterol.

Oils from fatty fish (herrings, mackerel, pilchards, salmon, and so on), still frequently denounced by nutritionists as contributing to obesity, may possibly have a greater protective effect than vegetable fats against heart disease. A preparation of concentrated fish oils (Maxepa) is available for those judged to be at special risk from heart attacks, but its protective value has not yet been fully confirmed.

Some foods, particularly egg yolks and fish roes, are particularly high in cholesterol but (contrary to earlier beliefs) do not appear, if not eaten in excess, to raise blood cholesterol levels any more than other animal fats. It is the total fat intake that matters.

A few people suffer from familial disorders, such as familial combined hyperlipidaemia, causing abnormal handling of fats in the body. This can lead to premature death from coronary artery disease unless they severely restrict their diet. Minor variants of this or similar disorders may be relatively common and may be a strong argument for restricting animal fat intake. However, things are less simple than they appear—in another rare familial disorder of fat handling blood cholesterol is at very high levels but often without any increased risk of heart disease.

In some disorders drugs may be needed to lower blood cholesterol levels, though they may have serious limitations because of their many side-effects such as nausea and other abdominal symptoms, itching or rashes, muscle weakness or pain, headache and sometimes impotence.

Fibre

Another factor which may lower blood cholesterol to some degree is fibre. In terms of diet, fibre does not include meat fibre. However tough this may be, it is ultimately digested. Dietary fibre is essentially indigestible and includes both obviously fibrous material such as bran and gum-like substances (long-chain polysaccharides). Among ordinary foods, peas and beans are major sources of fibre. Preparations of fibre such as bran, guar gum, and ispaghula husk, are available if there is too little fibre in the diet.

The effects of fibre are, first, to slow down starch and sugar absorption; second, to bind to and reduce absorption of fats; third, to act as a filler to lessen appetite; and fourth to give bulk to the diet. Bulk naturally helps to prevent constipation, by speeding up bowel movement. Rapid bowel movement may also help to lessen the absorption of toxic substances such as those which are thought to contribute to bowel cancer.

Fibres can have the unwanted effect of binding to and impairing the absorption of other substances. Guar gum, for example, significantly delays the absorption of penicillin from the digestive tract.

Fibre is particularly useful for diabetics (it may reduce their need for insulin by slowing the rise in blood sugar after meals), for the obese (provided that they do not indulge in too many baked beans which are high in fibre but full of sugar), and for some patients with bowel disorders (diverticulitis, Crohn's disease, and ulcerative colitis).

Protein

Many people are still obsessed with the idea that meat is essential for providing protein for 'body building', but there is no sound basis for this. Vegetable protein is just as effective—many North Indian races for example, are vegetarians but just as big and strong as steak-eating Texans. The only real value of meat protein is as a source of vitamin B_{12}. For this purpose it must be eaten almost raw but, as mentioned earlier, vegetarians rarely become vitamin B_{12} deficient.

Even milk is not essential for normal growth and babies thrive on vegetarian milk substitutes. For adults, milk may do more harm than good because of its fat content.

Vitamins

These catalysts of many body processes are essential only in small amounts but deficiency can cause serious disease or even death.

The trouble is that some people have got the idea that if a little is good, then a lot will be even better; others think that vitamins act in some mysterious way as tonics. This idea is happily promoted by, but is only of benefit to the manufacturers and health food shops—not the consumer. In actuality, vitamins are toxic in large doses and have, as a result of so-called megavitamin therapy, occasionally caused deaths.

Vitamins are *only* of value as medicines (as opposed to their normal intake in the diet) if genuine deficiency diseases develop. These can result from shortage of food (malnutrition) but in the Western world are rare and only likely to develop as a result of diseases affecting absorption. Infants occasionally become vitamin deficient as a result of some eccentric diet imposed by the mother. The remedy is of course to change the diet but if the mother persists in her prejudices then vitamins have to be given. Rarely, drugs can interact with vitamins.

Vitamin D is necessary for calcification of the bones. It is present in eggs and butter, abundant in fish liver, but is also synthesized in the skin when exposed to sunshine. Vitamin D deficiency is sometimes seen in Britain. The main group at risk are immigrants from the Indian subcontinent living in Northern Britain. Mothers and infants in particular may develop weak, poorly calcified bones (osteomalacia and rickets respectively). The effect of lack of sunshine may be compounded by eating chupatties made from wholemeal flour since this contains substances (phytates) which bind to calcium and interfere with its absorption.

Anxiety about vitamin D deficiency has lead mothers to give excessive amounts to infants, but this can be dangerous. If too much calcium accumulates in the blood, it can cause kidney damage among other effects.

Vitamin E was found to be essential for normal reproductive function in rats and many then assumed that it would confer increased sexual powers and various other mythical benefits on humans. Nevertheless there was for a long time no evidence that it was essential for humans and it has only recently been found that, in a few uncommon disorders of fat absorption, including some cases of coeliac disease, there is inadequate absorption of vitamin E. This can result in degenerative changes in the nervous system. Only in such cases (*not* sexual inadequacy) is vitamin E supplementation beneficial.

Vitamin C is present in fresh fruit and vegetables. Deficiency causes mood disturbances, rashes, and a bleeding disorder (scurvy). It is virtually unknown in Britain and many other parts of the world, as it requires many months of consumption of a highly abnormal diet to develop.

Vitamin A is also present in fruit and vegetables. Deficiency causes night-blindness and ultimately, damage to sight. In addition, the skin and mucous membranes become scaly (keratinized) and glands degenerate in a similar way. Again, it is difficult to become vitamin A deficient and overdose of vitamin A is poisonous.

In states of malnutrition, supplements of vitamin A and possibly of vitamin C may improve resistance to infection but there is no evidence that they confer any special benefits of this sort if the diet is adequate. Despite all the claims, there is little evidence that megadose vitamin C protects against the common cold. Vitamin A derivatives are also thought to have some anticancer effect and are useful for some skin diseases, but they are highly toxic. Some vitamin A derivatives such as etretinate, used for severe acne or psoriasis, are among the few drugs which are teratogenic and cause congenital abnormalities in human fetuses.

Vitamin B is a term used for a large group of substances with many different actions. Deficiencies of thiamine, riboflavin, and pyridoxine are hardly ever seen in Britain. Occasionally, this is a complication of chronic alcholism and the vitamins may then be given by injection. Pyridoxine deficiency can sometimes develop as a complication of administration of isoniazid for tuberculosis. Deficiences of several of the B group of vitamins can cause soreness and often atrophy of the surface of the tongue. Sore tongue is a common complaint but is hardly ever due to dietary vitamin deficiency and so not responsive to multivitamin tablets—though they have often been given and even more often taken without medical advice. Deficiency of thiamine results in neurological disorders and heart disease (beriberi). The cause of beriberi was discovered as a result of its appearance in the nineteenth century in the newly forming Japanese navy, where the sailors' diet consisted virtually only of polished rice and condensed milk—the disease was abolished simply by adding meat and vegetables to the diet.

Vitamin B_{12} and folic acid are essential for normal blood formation, as discussed later. There is some, but not as yet conclusive evidence, that some women need supplements of folic acid and possibly other B group vitamins during pregnancy to prevent congenital defects such as spina bifida in the fetus. Such vitamin supplementation should therefore be given as a precaution, to women who have had a baby with such a congenital defect, for any subsequent pregnancies.

It is worth repeating that 'low-grade' vitamin deficiencies do not develop in normal people on normal diets or cause people to feel weak, tired, or generally out of sorts. For such complaints vitamins do not act as tonics or pick-me-ups, except by self-suggestion. Multivitamin tablets are mainly of value to the manufacturers.

Minerals

Of the many minerals needed by the body, the only ones which may need to be given are iron (for normal blood formation), calcium, fluorides, or zinc.

Calcium is present in milk and other foods. It may occasionally need to be given with vitamin D when this is deficient. More controversial is its value in preventing or lessening weakening of the bones (osteoporosis) in later life.

Fluoride is necessary for reducing the risk of dental decay: calcium incidentally, is of no use for this purpose. Britain is behind most of the rest of the world in its reluctance to supplement the water supply with fluoride and the recent great decline in dental decay among children has almost certainly resulted from the general use of fluoridized toothpastes. Fluoride supplements can also be given to infants in tablet form. Those living in high-fluoride areas also have denser skeletal bone and may be less susceptible to osteoporosis in later life. Once osteoporosis has developed, however, fluorides are of no significant value.

Zinc is necessary for normal healing, and zinc supplementation may be necessary where there is severe loss as a result, for example, of large burns or failure of absorption due to intestinal disease. Deficiency is hardly ever the result of an inadequate diet.

Healthy eating: conclusions

Though the evidence for the role of the diet in many diseases, such as cancer or heart disease, is by no means conclusive, it seems sensible to follow the guidelines suggested by current research. Obesity should be prevented by avoidance of overeating and particularly of fatty foods such as butter, cream and fried foods. Additional fibre in the form of peas, beans and wholemeal bread, and plenty of fresh fruit and vegetables should be taken. Meat should be eaten in moderation and white meat such as chicken is preferable to red meat. Fish should be substituted for meat as frequently as possible. With such a diet or even with the more traditional British meat-and-two-veg, vitamin supplementation is unnecessary.

Weight reduction

Obesity is almost invariably the result of eating more than the body requires. Often it is the result of the wrong choice of food. Fried foods, for example, are flavoursome and easy to prepare but have a high caloric content.

Glandular and other causes of obesity are exceedingly rare and in any case are treatable only with reducing diets if the underlying cause cannot be remedied. It may be, as some evidence suggests, that the bodily (metabolic) processes in fat people are more efficient at converting food into fat, but as yet there is no way of changing a metabolic defect of this sort, if it is one. Reducing food intake is (unfortunately) the only remedy.

There are various approaches to the problem, but no satisfactory answers. The only truly effective way is to eat less, particularly of rich, high-calorie foods and, at least as important, to get more exercise. Eating less may be made easier by a high-fibre diet, by non-absorbable bulking agents (such as methylcellulose) to give a sense of satiety, or, transiently, by appetite suppressant drugs. Low-calorie diets are hardly ever completely satisfactory. Those who are overweight, are frequently those who enjoy food and for whom low-calorie diets lack the 'solidity' of real food and ultimately they relapse.

The food and related industries have no small share of the blame, in that we are constantly surrounded by advertisements for appetizing food and by restaurants emitting tempting aromas; cookery is one of the largest sections in many bookshops and domestic magazines are replete with recipes to add to the temptations.

Appetite suppressants are of very limited value. Some of the most effective (the amphetamines) are strong nervous system stimulants. Like cocaine, which has a similar action, they are strongly addictive and can cause psychoses in overdose. That aside, tolerance quickly develops to their appetite-suppressing action so that they become progressively less effective. Essentially similar drugs are diethylpropion and phentermine, which may have slightly less stimulant effects than amphetamines, but are also Controlled Drugs and lead to dependence. Fenfluramine is chemically similar to amphetamine but, by contrast, may be sedating or even cause depression. None of these drugs is particularly effective as, apart from the development of tolerance, appetite is only one stimulus to eating. Some people eat out of boredom, some because they are depressed, and many eat because they enjoy food.

Other drugs which may prevent obesity by such mechanisms as blocking carbohydrate or fat absorption from the digestive tract, are only in the relatively early stages of development and it is possible that adverse effects may outweigh any benefits.

Like giving up smoking or drink, restricting one's diet requires an iron will, removal of temptation, and something to encourage one to take more exercise—but who wants to walk or jog in dreary city streets, dodging the traffic, often in the rain and at the risk of being mugged? Even those who manage to lose weight, frequently relapse within a relatively short period.

The blood

Anaemia

The essential feature of anaemia is formation of too little haemoglobin (the substance which gives the colour to red blood cells and carries oxygen round the body) and this gives rise to many secondary effects. These, irrespective of the cause, include abnormal fatiguability, increased heart rate, breathlessness and, in extreme cases, heart failure.

The important blood-forming substances (iron, folate, and vitamin B_{12}) are present in a normal diet, so that deficiencies are, in affluent countries, more often due to poor absorption or to causes other than a defective diet.

Iron deficiency is the most common cause of anaemia, especially in women. The main cause is excessive blood loss as a result of heavy periods or childbirth. In men, iron deficiency anaemia is likely to be due to chronic blood loss from a peptic ulcer, cancer of the digestive tract, or piles. These can also of course affect women and the underlying cause of iron deficiency must always be investigated and, if possible, eliminated. Only then should iron (ferrous sulphate) tablets be given. Alternatives to ferrous sulphate are other salts such as ferrous fumarate which may be better tolerated.

Iron compounds were often (and perhaps still are) regarded as tonics but are of course only of value for iron deficiency anaemia. There are also many composite preparations of iron with additional drugs such as mixtures of vitamins and other minerals. These confer no additional benefit.

Vitamin B_{12} is unlike other vitamins in that, in Britain and comparable countries, deficiency is almost invariably the result of poor absorption due to atrophy of the stomach lining (pernicious anaemia). Less often, it is a result of intestinal disease. In most cases, vitamin B_{12} has therefore to be given by injection to ensure its absorption.

Those who eat no meat or any other animal products (vegans) might be expected to develop vitamin B_{12} deficiency but do so only exceedingly rarely. Under such circumstances, vitamin B_{12} can be given by mouth.

Folic acid is abundant in green vegetables and salads, but deficiency and anaemia can result from intestinal diseases which interfere with its absorption or, rarely, from taking drugs such as the anti-epileptic, phenytoin, which can occasionally interfere with its metabolism. If the underlying trouble cannot be dealt with or the drug changed, then folic acid should be taken as a medicine. This may also be necessary to meet the excessive demands for this important vitamin during pregnancy when additional iron may also be required. Iron and folic acid tablets are therefore useful for prevention of the common anaemia of pregnancy.

Anaemia as a side-effect of drug treatment

Anaemia can sometimes be a toxic effect of drugs, such as phenytoin. The antihypertensive drug, methyldopa, can sometimes cause anaemia as a result of a complex reaction which causes the red blood cells to burst (haemolytic anaemia).

Haemorrhagic diseases

Haemorrhagic diseases are of two main types: purpura, where purple patches due to bleeding into the skin are a major feature; and defects of clotting of the blood, as in haemophilia, where bleeding into deeper structures such as joints or the brain is the typical complication. Either type can cause prolonged bleeding after injuries or operations, but clotting defects are generally more serious.

Vitamin K is essential for normal clotting, another function of the blood. Deficiency under normal circumstances is rare, as vitamin K is formed by bacteria normally present in the gut. The newborn and particularly premature babies who lack these bacteria and have received too little from the mother, are given vitamin K_1 (phytomenadione) by injection to prevent haemorrhagic disease. Those with biliary or other diseases causing malabsorption of fats can also become deficient in vitamin K and for them a water-soluble preparation (menadiol) taken by mouth is recommended.

Vitamin K is only one of many substances necessary for normal blood clotting. Hereditary deficiencies such as haemophilia need to be treated by replacement of the missing factor (freeze-dried antihaemophilic factor) by injection. In severe cases this may need to be done regularly to prevent bleeding into body cavities or the brain as a result of minor injuries.

Drugs which affect haemostasis

Some drugs (the anticoagulants) are used deliberately to reduce the clotting ability of the blood and thereby lessen the risk of thromboses in veins. This is a hazard under such conditions as prolonged bed-rest, post-operatively or for some illnesses in the elderly. Heparin is given by injection for an immediate but brief effect, together with warfarin or an equivalent, by mouth: this has a delayed but prolonged effect by interfering with the formation of vitamin K-dependent clotting factors in the liver.

Some drugs can cause bleeding as an undesirable side-effect. Aspirin can increase this tendency by interfering with platelet function. Other drugs, particularly the antibiotic chloramphenicol, can sometimes poison the bone marrow and thus prevent platelet formation, causing bleeding into the skin (purpura) as a typical early sign.

Thrombolytic drugs (fibrinolytic agents) have been used to lessen the effects of coronary thromboses. Most of the drugs, such as anticoagulants, are of negligible value as they do not dissolve the already-formed thrombus. Recently, however, thrombolytic drugs (such as streptokinase and tissue plasminogen activator), have been shown greatly to increase survival from heart attacks if given within a few hours. Fibrinolytic drugs are also given for life-threatening thromboses blocking the veins of the lungs. They can cause severe allergic reactions.

Leukaemia

Leukaemia is mentioned here as an important, but uncommon, blood disease. It is not in any way related to nutrition but is a tumour of the white blood cells. These proliferate in the bone marrow at the expense of the red cells and platelets. Anaemia and bleeding tendencies are common results. Many abnormal white cells, unable to perform their normal defensive functions, are formed. Infections are therefore another important complication.

There are several types of leukaemia. Acute (lymphocytic) leukaemia, is one of the chief causes of childhood cancer but is potentially curable. Acute leukaemia (usually myeloblastic) in adults is quickly fatal in most cases. Chronic lymphocytic leukaemia, by contrast, tends to affect older adults and may be so slow in its progress that it may not need to be treated or have any noticeable effect on lifespan. Even if complications develop, people with this disease have been known to live into their 80s, but in others, progress of the disease may accelerate and be more quickly fatal.

Treatment of leukaemia is complex but in essence is by combinations of drugs (cytotoxic drugs) which in crude terms, are bone marrow poisons. They mostly have serious and unpleasant toxic effects, of which severe nausea is one of the hardest to bear. Radiation may be given, particularly to the brain and spinal cord, where drugs may be unable to reach the leukaemic cells. In simplified terms the effectiveness of these drugs depends on the fact that the leukaemic cells are slightly more sensitive to them than are the normal bone marrow cells.

Examples of cytoxic drugs given for leukaemia include cyclophosphamide, chlorambucil, doxorubicin, methotrexate, and cytarabine; these are discussed further in Chapter 13. Various combinations of these drugs are given for the different types of leukaemia. In addition, corticosteroids such as prednisolone in high doses, are valuable in lymphocytic leukaemia.

8

Respiratory disorders

The respiratory system includes the nose, throat, trachea
(windpipe), bronchial tubes, and lungs. The whole assembly
enables oxygen to be absorbed into the blood, and provides an
escape from the body for the waste gas, carbon dioxide. Since
we cannot survive without oxygen, serious diseases of the
respiratory system can be life-threatening. Respiratory diseases
are very common and varied in character—the main examples
are infections, such as the common cold, allergy, and tumours.
Smoking both increases susceptibility to, or worsens, virtually
all respiratory disease and is the main cause of cancer of the
lung.

The nose and nasal sinuses

The two most common conditions which affect the nose—hay
fever and the common cold—can both cause blockage of the nasal
passages. Although this is uncomfortable and inconvenient, nasal
blockage is not dangerous because you can breathe through the
mouth.

Hay fever

Hay fever is an allergic reaction to substances in the air in which
inhaling otherwise harmless particles, such as pollen, stimulates
the release in the body of histamine and other chemicals. These

chemicals cause the symptoms of hay fever, namely sneezing, nasal obstruction, and profuse watery secretions.

The term hay fever is a misnomer in that it can be triggered by many substances other than the grass pollen of hay, and there is no fever. The term allergic rhinitis (inflammation of the nasal mucous membrane) is a better description.

The ideal treatment is to avoid the allergen (the substance precipitating the allergic reaction). This is virtually impossible with some of the more common allergens, such as grass or tree pollens and the house dust mite, a microscopic insect, which is present wherever people live.

Antihistamines are a very commonly used type of drug treatment. They are usually fairly effective in blocking the actions of histamine in the nose but frequently also cause feelings of tiredness and sleepiness; hence these drugs must not be used by car drivers and operators of other dangerous machinery. Some of the newer antihistamines, such as astemizole (Hismanal), terfenadine (Triludan), and azatadine (Optimine), appear to cause less drowsiness. Antihistamines can also cause giddiness, bad temper, and a dry mouth.

An advantage of the antihistamines is that allergies are often multiple and various allergic conditions—in particular urticaria (hives) and other itching skin allergies—may be improved at the same time. Because the tablets are taken by mouth and are absorbed into the blood, they have a widespread effect. Paradoxically, antihistamines applied directly to the skin as a cream can cause allergic reactions when repeatedly used. In any case, this is less effective than taking antihistamines by mouth, even for skin conditions.

Sodium cromoglycate (Rynacrom) is inhaled directly into the nose as a powder or a liquid. Because this drug is not absorbed into the body, there are no generalized side-effects and over the years, sodium cromoglycate has proved to be remarkably safe. Cromoglycate acts by preventing the release of histamine and other allergy-producing chemicals within the body. It is quite effective, but, in severe hay fever, the nasal tissues are so swollen that it is difficult to get the drug into the nose.

Locally applied steroids such as a beclomethasone dipropionate (Beconase), betamethasone sodium phosphate (Betnesol), dexamethasone, budesonide (Rhinocort), and flunisolide (Syntaris) can be used for hay fever. Cortisone-like medicines (steroids) can produce widespread side-effects. However, when used in the treatment of hay fever, as nose drops or as a nasal spray, absorption into the body is slight. Their safety record is excellent and they are probably the most effective treatment for severe hay fever.

Decongestant nose drops (oxymetazoline (Afrazine), ephedrine, phenylephrine, oxymetazoline, xylometazoline (Otrivine)) are moderately effective and can be useful, but generally the longer they are used the less effective they become. Moreover, relief is sometimes only temporary and followed by a 'rebound' congestion of the nose, which is often worse than before. Persistent use of the drops can also cause permanent unpleasant and uncomfortable drying of the nose (atrophic rhinitis).

The common cold

The common cold produces some symptoms like hay fever. It is not one infection, but many with similar symptoms. The only too well-known inability of modern medicine to find a cure for common colds is due to the fact that they can be caused by many different viruses, and viruses are among the most difficult microbes to overcome.

In addition to the stuffy nose, there may be discomfort in the throat and general feelings of being unwell and miserable. Antibiotics have no action against viruses and there are as yet no safe and effective antiviral drugs for these infections.

So-called cold cures are not cures. They are usually mixtures which may contain two or more of the following:

1. Antihistamines, which dry secretions but usually increase feelings of sleepiness.

2. Decongestants such as phenylpropanolamine, phenylephrine, and pseudoephedrine, which may reduce swelling of
   the linings of the nasal passages, but may increase the blood
   pressure or heart rate.

3. Cough suppressants such as codeine, pholcodine, and
   dextromethorphan. These can depress breathing and make
   removal of sputum difficult. Both of these actions are
   dangerous for patients with asthma or bronchitis.

4. Pain-killers such as paracetamol, aspirin, and codeine. These
   may lessen some of the unpleasant local and general
   symptoms of a cold but do not help to overcome it.

5. Belladonna derivatives such as atropine and hyoscine can
   diminish watery secretion from the nose, but may also
   produce mental changes— atropine stimulates the brain but
   hyoscine causes drowsiness.

So far, therefore, we have no drugs to cure or greatly ameliorate
the common cold, and many of the drugs which may be used can
cause undesirable side-effects.

The effect of vitamin C is still unproven, but may play a minor
role in reducing the frequency and severity of colds. If you think it
helps by all means take it—but not too much, one person (at least)
has died from an overdose of vitamin C.

Sinusitis

The sinuses are spaces which communicate with the nasal cavity
within the bones of the face. They are usually affected at the same
time as the nose in nasal allergies and in the common cold.
Sometimes the sinuses become blocked by swelling of their lining.
If bacteria infect them, then pain (sometimes intense) in the head
and face can be caused by the formation of pus in the sinuses and a
build-up of pressure. When the blockage is relieved by shrinking
the lining mucous membrane, pus may escape through the nose.
This is accompanied by rapid resolution of the pain and infection.
The condition can begin suddenly and be relatively brief (acute

sinusitis) or persist for long periods (chronic sinusitis) with dull and intermittent pain.

The aim of medical treatment is to shrink the sinus lining to allow drainage and resolution. Decongestant nose drops (such as ephedrine) may help—but one of the more effective remedies is inhalation of steam. Aromatic oils can be added to the hot water, but the steam itself is the more active ingredient. The only hazard of this treatment is spilling the jug of hot water.

Antibiotics frequently fail to give any relief. Either the bacteria may not be vulnerable to such drugs or the primary cause, namely the swollen mucous membrane, continues to obstruct the outlet of the sinuses into the nose.

The pain of sinusitis can be severe enough to justify the use of pain-killers. Aspirin may be useful—not only does it relieve pain but also reduces the swelling of inflamed tissues. Paracetamol or codeine are alternatives which can be used by people with indigestion or gastric ulcer.

Diphtheria

Diphtheria is a rare disease in Britain and remains so as a result of routine childhood immunization (vaccination). Diphtheria causes a mild sore throat but considerable swelling of the lymph nodes ('glands') in the neck. The bacteria responsible colonize the throat causing a sloughing ulcer, but the disease results mainly from the toxin that the bacteria release from the throat into the bloodstream. This toxin can cause paralysis or death from its effect on the heart. In young children the diphtheritic slough ('membrane') can extend down to block the narrow entrance to the airway and cause asphyxia.

Though the diphtheria bacteria are sensitive to antibiotics, such as penicillin, this treatment does not combat any toxin that is released. Antitoxin must therefore be given. In addition, in infants, it may be necessary to open the airway by putting a tube into the throat (tracheostomy) to enable them to breathe.

More important is to prevent diphtheria. This can be effectively and safely done by immunization. Where immunization has stopped, either because of over-confidence that diphtheria has disappeared or because of anxiety about imagined dangers of immunization, epidemics of this infection have broken out. (This happened recently in Texas and in West Germany.) In the underdeveloped parts of the world, diphtheria is common and dangerous because of the lack of widespread immunization.

The lungs

Influenza

Influenza is another viral infection for which there is no specific treatment. It usually causes a fever, aching in the muscles and other symptoms which can be lessened to some extent by aspirin or paracetamol. There are, however, very many strains of influenza virus and the illness can be mild or a killing disease, as in the case of the 'Spanish' influenza epidemic of 1918–19 which killed tens of millions.

Amantadine is an antiviral drug which gives some protection against A strains of influenza viruses and may sometimes be helpful for the elderly who are at risk from various complications of this infection. Amantadine is only useful in prophylaxis, however, and is of no use once the infection is established. It can produce unpleasant side-effects such as nightmares and feelings of nervousness.

For those in whom an attack of influenza might have serious consequences, such as patients with severe bronchitis, influenza vaccination may be used to reduce the chance of an attack. The vaccine itself can cause allergic reactions and is of no use if an outbreak is due to a new strain of the virus.

Chronic bronchitis

Chronic bronchitis is a persistent state of inflammation of the bronchial tubes causing cough and sputum production. The condition tends to flare up repeatedly—usually after a cold or similar infection in the winter. The patient is often breathless due to the air sacs in the lungs becoming damaged and over-distended (emphysema). In severe emphysema, the patient can become barrel-chested as a result of inability to exhale fully from the inelastic lungs, and breathless and blue from lack of oxygen.

Chronic bronchitis may be caused by smoking, an inherited predisposition, asthma, and atmospheric pollution. Cigarette smoking always results in some degree of bronchitis.

Ideally, bronchitis should be prevented—particularly by not smoking. Once the disease is established, treatment can do little more than slow its progress. Thus treatment is aimed at avoiding aggravating factors such as smoking, polluted atmospheres, or cold bedrooms.

Chest infections aggravating chronic bronchitis should be prevented or treated where possible, for example, by influenza vaccination or antibiotics for bacterial infections. It may not be possible to relieve breathlessness due to severe emphysema—and physiotherapy to encourage increased respiratory effort may be needed. Suppressing the sensation of breathlessness with sedatives or cough suppressants is dangerous and can be fatal.

Bronchodilator drugs Many patients with bronchitis who are breathless and wheezy have an asthma-like obstruction of the smallest divisions of the bronchial structure (bronchioles), particularly if they are prone to allergy. This may be relieved by bronchodilator drugs, such as salbutamol (Ventolin) or theophylline.

Expectorants are intended to thin and loosen sticky sputum to enable it to be coughed up from the bronchi. No drugs have really been shown to have this desirable effect. As yet therefore, there seems to be no more effective method of removing sputum than active coughing and physiotherapy. The latter includes tipping the patient's head down (postural drainage) and pummelling the chest. Other cough mixtures, misleadingly called expectorants, contain antihistamines or codeine, and suppress cough. This may be comforting, but can be harmful in conditions such as chronic bronchitis.

Antibiotics Apart from treating acute flare-ups of infection, antibiotics do not seem to help in the long-term treatment of bronchitis.

Asthma

Asthma is a disease in which there are intermittent (reversible) attacks of wheeziness (readily audible to another person) and breathlessness which is frequently, but by no means invariably, due to allergy. Between the attacks, the patient may be comfortable and able to breathe perfectly normally.

Allergic asthma is one of a group of diseases in which susceptible persons have unpleasant reactions to a great variety of substances (allergens) such as pollen grains, animal hair (dander), some drugs (particularly penicillin), and various components of house dust, which are completely harmless to other people. The allergy-prone person produces antibodies to these substances but these antibodies seem to lack their normal protective function and instead trigger the release, from cells in the patient's body, of histamine and other chemicals. These internal chemicals cause the airways to become obstructed mainly as a result of spasm, mucus production, and inflammatory swelling of their linings.

Allergy is only one of several mechanisms causing asthma. In many other sufferers, this spasmodic obstruction of the airways can be precipitated by infections (colds, flu, etc.), cold air, exercise, emotion, or even laughing. In all patients with asthma, the airways appear to be 'irritable'; in other words, hyperresponsive to various stimuli.

The main strategies in the treatment of asthma are, wherever possible, preventive. However, it is often impossible to avoid such triggering factors as pollens and in some patients there is no discernible precipitating cause to avoid. The main lines of treatment are then to block the process leading to airway obstruction at some stage with various drugs.

Sodium cromoglycate (Intal), which is inhaled as a powder or spray, is often very effective in preventing the release of histamine and other chemicals in the body. Cromoglycate is taken regularly two or four times a day and, in children, particularly, it often greatly reduces the frequency and severity of asthma brought on by allergy or exercise, but not when associated with infection. It is, however, of no use during the attack.

Bronchodilators form the main basis of day-to-day treatment of asthma as they relax the spasm and reduce swelling and mucous plugging in the fine bronchial divisions of the lungs. Older drugs of this sort such as adrenaline were given by injection, or others such as ephedrine by mouth. The disadvantage of these routes of administration is that these drugs, which have multiple effects, enter the circulation and reach all parts of the body. Unpleasant or dangerous effects on other organs—such as the heart—could result.

A great advance was the development of aerosols which could be inhaled directly into the lungs so that effects on other parts of the body were reduced. This allowed much lower doses of drugs such as isoprenaline to be used. A further advance was the development of drugs which have a powerful but selective action in dilating the bronchial tubes but little or no effect on the heart. Safety and efficacy are also increased by giving these agents as an inhaled aerosol. Examples are pirbuterol (Exirel); salbutamol (Ventolin), and rimiterol (Pulmadil). In particularly bad attacks of asthma, bronchodilators such as salbutamol or aminophylline may be given by injection.

In some patients, asthma does not respond to bronchodilator drugs. In such cases, steroids (short for corticosteroids) can be valuable and even life-saving. Many people are anxious about the idea of using these steroids, which are often referred to as 'cortisone'. This is because corticosteroids given for a long period for any of a great variety of diseases can cause serious side-effects, such as increased susceptibility to infection. The important adverse effects of corticosteroids are discussed in more detail in chapter 10, but it must be appreciated that the body itself also produces corticosteroids and life depends on it doing so.

In asthma, corticosteroids such as beclomethasone dipropionate (Becotide) or betamethasone valerate (Bextasol) are usually given as bronchodilators and inhaled as aerosols into the airways. By being brought into direct contact with the malfunctioning bronchioles, microscopic doses of these corticosteroids are effective, and since the rest of the body is hardly affected they are very safe. Their main side-effect, seen in only a few users, is the development of thrush in the mouth and throat, but this is not dangerous.

In the most severe cases of asthma unresponsive to other forms of treatment, corticosteroids have to be given by mouth in larger doses. This has concomitant side-effects of which plumpening of the face and reddening of the complexion are the most obvious (see also Chapter 10). Despite these very undesirable side-effects, corticosteroids are essential treatment under expert medical supervision. A significant number of patients die from asthma as a result of under-treatment.

Whooping cough

As many parents know, whooping cough is a feverish infection resembling a cold at first but going on to violently severe paroxysms of coughing which cause the characteristic 'whoop' as the child tries to regain breath. Thick mucus is coughed up and vomiting may be associated. Inability to breathe during the coughing attacks can lead to asphyxia. This in turn, by reducing the blood supply to the brain, may cause fits and brain damage. The sheer violence of the coughing can even result in bleeding into the brain by forcing up the blood pressure.

Although the antibiotic, erythromycin, is effective against the bacterium causing whooping cough and shortens the duration of the illness, it is of little value in the acute paroxysms of coughing. Prevention of this potentially lethal disease is therefore essential. But here, alas, arises an example of one of the great current medical controversies—the vaccine can *very occasionally* cause brain damage (encephalitis). Very many parents have therefore refused whooping cough vaccination for their children. As a consequence there has been a dramatic rise in both the incidence of and mortality from this disease.

It has taken a long time to establish with certainty how great or small the risks of whooping cough vaccine are. Part of the trouble was caused by earlier vaccines being less pure than the present ones. Also, some of the cases of encephalitis that had been ascribed to the vaccine were due to coincidental causes. Even if the risk of encephalitis from whooping cough vaccine is only one in a million it is no comfort to the parents of that millionth child who gets permanent brain damage. However, there is no doubt that the risk of brain damage or other complications are far greater from whooping cough than from the vaccine. Vaccination is a valuable protection. No doctor can advise as to whether any given infant is particularly likely to be the rare unlucky one, except that the vaccine should not be given to infants who have previously had fits or convulsions or to those who have any significant reaction to the first dose of the vaccine.

The whole problem boils down to how one perceives the levels of particular risks. Of course, for example, we are all justifiably worried about terrorist bombs, but how many seriously consider the far greater risks of driving on a motorway? Similar considerations apply to the perception of risks from drugs such as vaccines.

Pneumonia

Pneumonia is a severe infection of the lungs which extends to the finest divisions of the bronchial system. Untreated pneumonia is a serious disease with a high mortality rate, particularly in the elderly and debilitated. In the past, a common form of this infection (lobar pneumonia) attacked the young and in these too was often fatal. The conquest of lobar pneumonia was one of the first triumphs of the sulphonamides ('sulpha' drugs) which were the forerunners of the antibiotics.

One physician has graphically described how in the 1930s there would be many patients with lobar pneumonia in the wards and though some survived with the help of their own body's defences, many inevitably died. Then, quite suddenly, as a result of giving sulphonamides, patients survived against all expectations and, to the amazement of the doctors and nurses, were converted overnight into comfortable convalescents.

The antibiotics, including penicillin, have replaced the sulphonamides for pneumonia. Antibiotics are, however, ineffective for viral pneumonia caused, for example, by the influenza virus and as with so many other viral infections, there is as yet no effective treatment.

Patients with AIDS frequently die from pneumonia due to the parasite *Pneumocystis carinii* which is completely harmless to normal people.

When pneumonia involves the surface covering of the lungs (the pleura) this also becomes inflamed (pleurisy). The result is severe pain in the chest over the involved part. The pain is aggravated by taking a deep breath and by coughing—just the things which the patient with pneumonia would be inclined to do. Pain-killers such as codeine, or even morphine, may be needed.

Pulmonary tuberculosis

This is an infection of the lungs by the tubercle bacillus which causes a chronic cough, weakness, loss of weight, and often coughing up of blood (haemoptysis). Tuberculosis (consumption) was justifiably called one of the Captains of the Legions of Death and world-wide is still a cause of a vast mortality.

Tuberculosis used to be treated by prolonged bed-rest, a highly nutritious diet, fresh mountain air, repeated X-rays and sometimes operations to collapse the affected lung in special hospitals (sanatoria). All this was quite ineffective—only drugs can reliably overcome the tubercle bacillus once it has gained a hold.

Nevertheless, even before the introduction of the first antituberculous antibiotic (streptomycin) in the late 1940s, tuberculosis was a declining disease probably largely because of better living conditions and control of the sources of infection which reduced the spread of the disease. Antibiotics, however, have had so dramatic an effect on the mortality that the building of sanatoria became unnecessary in the early days of the National Health Service.

Among immigrants from the Indian subcontinent tuberculosis is at least 50 times more common than in native Britons. Paradoxically, tuberculosis has become so unusual in the native British population that it is rarely suspected and the death rate is relatively higher. Usually the disease affects those in later life and another contributory factor may be that some of them are vagrants who fail to seek medical care or fail to take their antibiotics regularly.

Modern treatment of tuberculosis is with rifampicin, ethambutol, isoniazid, and pyrazinamide. A cocktail of these antibiotics is used because the tubercle bacillus is able to develop resistance more quickly to a single antibiotic.

These are very potent antibiotics and will often even enable patients with AIDS to recover from tuberculosis, to which they are very susceptible. Inevitably such drugs can have serious side-effects. Streptomycin, the original antituberculous drug, though effective, is usually avoided because it can cause permanent deafness. Another now obsolete form of streptomycin could cause irretrievable loss of balance. Disasters such as these mainly resulted from prolonged use.

Rifampicin can disturb liver function and can sometimes cause influenza-like symptoms, abdominal or respiratory disturbances, and damage to the kidneys. The drug can reduce the effectiveness of oral contraceptives. Of no great significance but possibly alarming to the patient is that rifampicin can colour the saliva and urine orange. Side-effects of ethambutol are rare, but it may cause disturbances of vision. Pyrazinamide can occasionally cause liver damage, while isoniazid can cause nerve lesions, nausea, and allergic reactions such as rashes.

These undesirable effects of antituberculous drugs are by no means inevitable consequences of their use and with careful monitoring of the patient, and avoiding overlong use of very high doses, they can often be avoided. Moreover, if such troubles as these develop, it is feasible to switch to another antibiotic and many patients are cured of tuberculosis without experiencing serious side-effects. Only patients who are a potential source of infection need remain in hospital. A patient can usually safely go home even after the first six weeks of treatment; the drugs are then continued for a further 7 months or less.

Fibrocystic disease (cystic fibrosis)

This disease is inherited from both parents as a recessive trait and the abnormal gene is usually carried by both parents. The basic defect involves glands: sweat, digestive secretions, and mucous glands in the lungs are all abnormal. In the digestive system, the pancreas is particularly blocked and its secretions cannot reach the intestine to digest fat and other substances. However, the main problem is in the lungs. Here, thick plugs of mucus block segments of the lung and can cause areas of collapse and infection. Antibiotics are certainly valuable in treating the many episodes of infection, but do not remove the primary cause of the lung condition, mucous plugging.

The disease is tragic for both the affected children and their parents. The children are usually stunted, constantly ill and need repeated sessions of physiotherapy to help bring up the abundant thick mucopurulent sputum.

In fibrocystic disease, mucolytics such as carbocisteine (Mucodyne), and methylcysteine (Visclair) were thought to help in dissolving the thick mucous plugs, which are an especially severe problem in the small airways of children. Though many have great faith in these drugs, there is little objective evidence that they are of any significant value.

But it is easy to understand why people clutch at anything that might help with this dreadful problem.

Legionnaires' disease

This disease is an increasing cause for concern. It acquired its name, not from any association with the armies of ancient Rome, but from its recognition after an outbreak among American Legion ex-servicemen who had attended a meeting in Philadelphia in 1976. Legionnaires' disease is unusual in that the infection does not seem to spread by any of the usual routes but mainly in aerosols of water, particularly from air-conditioning plants. In a recent outbreak, even passers-by in the street were affected. The source of the bacteria and how they got into the water is a mystery. This makes prevention difficult as it is hardly feasible to test every air-conditioning plant, just in case it harbours the bacteria. In some cases in which the organism has been found, no infections had apparently resulted. In other cases, elimination of the microbe has proved so difficult that there have been recurrent outbreaks of the disease.

Many, particularly the young, acquire this infection without any ill-effects. Others, particularly the elderly and those with pre-existing respiratory disease, develop pneumonia which can be fatal. It seems possible that some of the fatal pneumonias of unknown cause among hospital patients have been due to this infection. A febrile illness without pneumonia (Pontiac fever) is another form of the disease.

Early recognition of the cause of the infection is essential, as it should then respond to the antibiotics erythromycin or rifampicin.

9

Reproductive and urinary disorders

The urinary and reproductive systems have separate functions. The secretion of urine is essential for the maintenance of the chemical equilibrium of the body. This is quite distinct from the organs concerned with reproduction. However, the two systems are often considered together (the genito-urinary system) because of their anatomical proximity.

The urinary tract

This consists of the kidneys and bladder, and the tubes which connect them and also conduct the urine from the bladder to the outside.

The kidneys play a vital role in controlling the concentrations of many of the chemicals in the blood. Without the normal activity of the kidneys, urine is not produced and the patient dies having suffered from severe chemical disturbances. However, the majority of diseases of the urinary system are not life-threatening and readily resolve.

Acute (short-term) infections

Cystitis and pyelitis

Cystitis and pyelitis are infections of the bladder and outlet of the kidney, respectively. They often co-exist. The patient feels ill, may have a fever, and experiences discomfort on urinating. Repeated attacks are common. Such infections are more common in females than in males. This appears to be because the bacteria enter the bladder from the outside, and the much shorter passage which drains the bladder in the female provides a less efficient barrier than in the male.

If the precipitating factors can be identified, then it may be possible to prevent recurrences. Simple measures such as increasing the intake of liquids or urinating after intercourse may be sufficient. The use of a bidet and wearing porous cotton underwear can also help.

In a severe attack, the patient feels ill and should be nursed in bed. Six or more pints of liquid should be drunk daily. The doctor may send a specimen of urine to a laboratory to find out which organism is present and which antibacterial drugs are likely to be effective. Whilst waiting for the result of these tests, drug treatment may be started—the agents most likely to be appropriate are trimethoprim and amoxycillin. However, sometimes the results of bacterial culture may necessitate a change in drug treatment.

Non-specific urethritis

The urethra is the passage connecting the bladder to the exterior. A discharge of pus from the urethra is due to an infection, which in at least half the cases is caused by chlamydia, a micro-organism. In the UK, chlamydial urethritis is the most common sexually transmitted disease. It can lead to complications which include inflammation of the cervix and uterine tubes.

The condition is responsive to treatment with an antibiotic, usually oxytetracycline given for three weeks.

Gonorrhoea

This presents in men as urethral discharge and a burning sensation on passing urine. In women there is frequent and painful urination and a vaginal discharge. The infection is usually venereally acquired. In the past, infertility, urinary obstruction, and infection of joints followed the onset of gonorrhoeal urethritis, but modern treatments have made these complications rare in the UK. A single large injection of procaine penicillin is usually adequate treatment. For those patients who are allergic to penicillin, a single injection of spectinomycin or oral co-trimoxazole for four days are effective alternatives.

Syphilis

This venereal disease is a serious infection which can damage almost any organ in the body. The heart, large blood vessels, and brain are particularly at risk.

Since the introduction of penicillin, the incidence and hence spread of the disease have fallen dramatically, and in individual patients the condition can be halted if treatment is begun at an early stage of the infection. If a pregnant woman has syphilis the unborn child may die in the uterus or may be born alive but suffering from the infection (congenital syphilis). The eyes, teeth, bones, and brain may be affected.

Syphilis in its early stages can be cured by daily injections of procaine penicillin for ten days. Late syphilis requires three weeks of daily penicillin injections. Congenital syphilis can also be arrested by penicillin injections, but is best prevented by testing pregnant women for the presence of the illness and, if necessary, giving penicillin treatment during pregnancy.

Genital herpes

Several types of herpes virus can infect humans. Types 1 and 2 of the herpes simplex virus (HSV 1 and HSV 2) are spread by sexual intercourse. The usual cause of genital herpes is HSV 2.

The local lesions on or near the genitalia caused by HSV 2 start as red, painful, or itching areas which become small blisters. These burst to produce painful ulcers. Recurrent attacks tend to be less severe, but the first episode may be accompanied by fever, headache, and difficulty on urinating.

The local discomfort can be eased by the application of an anaesthetic ointment, cream, or jelly. Such agents include lignocaine (Xylocaine), benzocaine (Solarcaine), and amethocaine

(Anethaine; Locan). However, repeated use of local anaesthetics on the skin can cause an allergic reaction.

Severe symptoms can be reduced and healing promoted by intravenous or oral acyclovir. However, this does not prevent recurrences.

Chronic (long-term) infections

Chronic pyelonephritis

This is a long-term infection of the substance of the kidney by micro-organisms which stimulate the formation of pus and destroy the kidney tissue. Why this should happen is a mystery because the great majority of patients who suffer acute urinary infections (such as pyelitis and cystitis) recover completely, and even recurrent attacks very rarely cause any lasting damage to the kidneys.

In chronic pyelonephritis, if large amounts of kidney tissue are destroyed, kidney function is impaired. This means that the control of the chemical content of the blood plasma (which the kidney normally exerts) is lost. The chemical disturbances which result cause disease of bones, vomiting and headache, and can lead to coma and death. The blood pressure may rise to such an extent that it causes fits. Another common cause of chronic renal failure is glomerulonephritis, which requires similar treatment.

The use of an artificial kidney (renal dialysis) can rapidly reverse these changes. However, the machines are expensive to buy, use, and maintain. The major difficulty is that for the rest of the person's life, the blood has to be circulated through the machine (which process takes several hours) on two or three occasions each week. The alternative is to find a compatible donor kidney. The donated kidney is transplanted into the patient's pelvis and functions like a normal kidney. The only way a completely compatible kidney can be found is if the patient is lucky enough to have an identical twin who is willing to donate a kidney. Failing this the match will not be perfect and drugs will have to be taken to prevent the donor kidney from being rejected. Such drugs are called immunosuppressive agents and include prednisolone and cyclosporin. Prednisolone is a steroid—its side-effects are described in Chapter 10. Cyclosporin inhibits the cells which cause graft rejection. It is a potentially toxic drug—large doses can damage the kidney. Even correct doses can cause nausea, vomiting, feelings of misery, swelling of the gums, and excessive growth of hair.

The reproductive system

Oral contraceptives (in the female)

The most important drug which influences the reproductive system is the contraceptive pill. There are two types: the combined pill and the 'mini'-pill.

The combined oral contraceptive

This contains two hormones: an oestrogen and a progestogen. This mixture results in a powerful antifertility action because several reproductive functions are altered simultaneously: ovulation is prevented, the entrance to the uterus is made impermeable to sperms, and the uterus does not allow a fertilized egg to remain in its cavity.

This preparation is taken daily for 21 consecutive days. Medication is then stopped for 7 days, during which time uterine bleeding begins 2–3 days after the last tablet was taken.

Considering the large numbers of such tablets that are being consumed, serious toxic effects are very rare. However, a wide variety of problems of varying magnitude have been reported. They can be classified according to the component of the pill thought to be responsible.

Toxicity due to oestrogen:
    uterine bleeding in the middle of the cycle;

aggravation of diabetes;
risk of blood vessel disease, such as stroke, heart attack, leg vein thrombosis, blood clots in the lung;
yellow patches in the skin of the face;
jaundice;
migraine and other forms of headache;
swelling of the legs;
vaginal discharge.

Toxicity due to progestogen:
acne;
increased hair on the face and trunk;
vaginal dryness.

Toxicity due to both hormones:
depression;
raised blood pressure;
failure to resume normal menstruation on stopping the pill;
breast swelling.

To put the most serious of these problems into perspective, it has been estimated that pill users have a risk of 1 in 5000 of dying of cardiovascular disease each year of pill use—but the cases of this are mainly concentrated in older women (over 35 years), those with raised blood pressure, and smokers. Also this risk must be compared with the natural risks of pregnancy, which in this age group has a mortality of 50 per 100 000 (1 in 2000). Mortality rates for disease due to blood clots in the arteries and veins of women aged 35–44 years are as follows:

pill users	3.9/100 000
non-users, non-pregnant	0.5/100 000
during pregnancy	2.3/100 000

Other hazards of the pill include an increased risk of gallstones, and cancer of the uterus and cervix (see Chapter 10). However, the higher incidence of cancer of the cervix may be due to prolonged sexual activity and not be directly related to medication. By contrast, there is a decreased incidence of benign breast lesions and of ovarian cysts. There is at present controversy as to whether the pill significantly affects the incidence of breast cancer.

Despite the low incidence of serious toxic effects, there are certain circumstances in which the pill should not be used. These include:

active diseases of the veins or arteries;
high levels of blood cholesterol and other lipids;
severe liver disease;
cancer of the breast or uterus;
heart disease;
high blood pressure;
diabetes which is not readily controlled.

Low-dose progestogen-only preparations ('mini-pill')

These may be used in slim patients if the combined pill is poorly tolerated or contra-indicated. The dose per Kg bodyweight is too low in obese women; they get pregnant on the low-dose pills. The chance of becoming pregnant on this low-dose pill is higher than with the combined pill. Pregnancy rates per 100 women per year are:

barrier methods	1.5–2
intra-uterine contraceptive device	1.5–2
low-dose progestogen pill	1.5–2
combined pill	0.3

The other problems of the 'mini-pill' are irregular uterine bleeding, breast tenderness, skin flushing, acne, and headache.

Other uses of sex hormones

The main non-contraceptive uses of sex hormones are in the field of gynaecology. Oestrogens and progestogens are used to treat abnormalities of menstruation. Oestrogens are also effective in preventing thinning of bones (osteoporosis) after the menopause.

Infertility in the female can be treated with hormones or related substances (such as clomiphene) only if the condition is due to a hormonal lack which has caused a failure to ovulate.

Sex hormone replacement therapy

The female menopause may be characterized by swings in mood, insomnia, nervousness, loss of confidence, and headaches. These do not usually persist, and an understanding of their cause can help to reassure the sufferer that these are common self-limiting phenomena. More severe cases may be helped by oestrogen treatment.

The menopause is caused by a reduction in the activity of the ovary, which secretes sex hormones such as oestrogens. However, giving oestrogens does not relieve all the uncomfortable features of the menopause. Those which are helped by oestrogens are hot flushing, genital tract thinning and drying, and bone pain due to bone thinning (osteoporosis).

The oestrogens most commonly used for the treatment of menopausal flushes are stilboestrol and ethinyloestradiol. Nevertheless, these drugs are avoided whenever possible, and if they are used, the least dose is given for the shortest time.

There is no doubt that post-menopausal osteoporosis is prevented by oestrogens. However, these can predispose to arterial disease and to blood clotting and there is also an increased risk of cancer of the uterus. The risk to the uterus may possibly be reduced or eliminated by the simultaneous administration of a progestogen such as norethisterone. Another problem is that even if oestrogen replacement does prevent bone disease, the condition will develop as soon as the drug is stopped. It does not seem reasonable to continue treatment with this hormone for the rest of a person's life.

Thinning, drying, and tenderness of the female genital organs after the menopause can safely be treated with a local application of an oestrogen cream to the affected parts. This does not carry with it any known risk to the uterus or blood vessels.

If there is a male menopause it does not seem to be due to lack of the male sex hormone (testosterone). However, deficiency of the male hormone due to disease or absence of the testes or due to pituitary gland disease can be effectively treated with hormone replacement. Impotence in middle-aged and elderly men does not usually respond to testosterone. Male sex hormones in large doses increase the bulk and power of muscles. However, the blood cholesterol is raised at the same time and the chance of developing arterial disease and heart attacks is increased.

10

Disorders of hormones and endocrine glands

A gland is a part of the body which produces specialized chemical substances. Examples are the salivary glands which secrete into the mouth, intestinal glands which secrete into the intestine, and the breast which secretes milk to the outside of the body. The glands which secrete chemicals directly into the blood are called endocrine glands. The substances they provide are called hormones. Diseases can be the result of deficiencies or excesses of hormone secretion of a gland.

The thyroid

The thyroid is an endocrine gland situated in the neck. One of its main secretions is the hormone thyroxine. Its function in the body is the stimulation of cells to produce heat, stimulation of chemical metabolism, and, in children, it contributes to growth and development. All of these effects are due to direct actions of the thyroid hormone on the cells of the body.

Thyrotoxicosis

This is a disease due to excessive secretion of the thyroid hormones. The thyroid gland may be obviously enlarged (goitre) but this is not invariable. The main consequences are due to

137

stimulation of metabolism. Thus, the patients find they cannot tolerate a warm environment, the appetite is stimulated but weight loss is common; the heart beats more quickly and sometimes develops an irregular rhythm, and the eyes may protrude and become prominent.

The choice of treatment lies between drugs, an operation, or shrinking of the gland by exposure to radioactive iodine.

The simplest treatment is to take an antithyroid drug such as carbimazole or propylthiouracil. These act directly on the cells of the thyroid gland and stop the chemical processes involved in the manufacture of the thyroid hormones. These drugs do not act immediately but take 1–2 months to reduce thyroid activity to normal. The drug treatment is continued for 12 to 18 months. Many patients remain well afterwards but within another two years at least half of the patients will have relapsed.

An operation for thyrotoxicosis consists of removing part of the thyroid gland. This is usually successful, but sometimes too much or too little of the gland is removed which results in over- or under-correction.

Radioactive iodine is usually reserved for patients over the age of 40 years. This is a precaution in case this treatment leads to a delayed toxic effect. In fact, no such delayed actions have been observed. The radioactive iodine (^{131}I) is given in the form of colourless, tasteless drink, usually taken via a straw. The main difficulty with this treatment is that the shrinking of the gland which follows, may progress too far and cause under-functioning of the gland.

Hypothyroidism (myxoedema)

This is due to under-functioning of the thyroid gland. The patient becomes physically sluggish, over-sensitive to the cold, develops a husky voice, and may notice pallor and thickening of the skin. The condition is excellently treated by giving thyroxine sodium by mouth. The treatment is usually lifelong.

Hypothyroidism from birth is the cause of cretinism. Without treatment it results in severe mental deficiency and dwarfism. Treatment with thyroxine sodium from the first month or two of life allows normal physical growth and mental development.

The adrenal cortex

The adrenal gland secretes several hormones. These include adrenaline, aldosterone, sex hormones, and cortisol. The last three are steroids and are produced in the outer layers of the gland (the adrenal cortex) and adrenaline is made in the inner core (the adrenal medulla). The adrenal cortex is essential for life. If this part of the gland has been destroyed by disease or removed at operation then cortisol, or a similar substance, has to be given for the rest of the patient's life. This type of treatment is called replacement therapy.

The group of abnormalities due to under-activity of the adrenal cortex is called Addison's disease. This includes weakness, low blood glucose, increased pigmentation of the skin and mucous membranes, and low blood pressure. The patient eventually becomes extremely weak, collapses, and dies.

The opposite situation, over-secretion by the adrenal cortex causes accumulation of fat in the trunk, weak bones, a swollen congested face, and diabetes. This conglomeration of signs is called Cushing's syndrome.

In the treatment of Addison's disease cortisol may be all that is required to maintain health by replacing the missing hormones. The normal adrenal cortex secretes about 25–50 mg of cortisol during a non-stressful day. However, if there is pain, disease, injury, excessive heat or cold, or emotional stress, the body adapts by increasing cortisol secretion up to four or five fold. Thus, for the patient with Addison's disease the usual daily dose is 50 mg of cortisol, but if a serious infection develops or an operation is carried out, the daily dose may have to be increased to 200 mg. If this adjustment is not made, then the patient will not be able to withstand the stress and will develop the features of adrenal failure; this can be fatal.

The above sketch of replacement of cortisol in Addison's disease relates in a reasonably logical way to what is known of the normal functioning of the adrenal cortex. However, when used in much bigger doses than those discussed, cortisol shows some quite strange and unexpected properties. These include reduction or even complete suppression of the signs and symptoms of inflammation. Thus, many medical conditions can be altered because the features of infections, trauma, and many rheumatic diseases are due to inflammation. Nevertheless, the underlying disease is not removed, and may even worsen because inflammation has a useful role in combating infections. Another drawback is that such large doses of cortisol and other steroids lead to a clinical picture identical to that of over-activity of the adrenal cortex (Cushing's syndrome). This means that these patients have plump pink faces, thinning of the bones, fat deposition on the trunk, and stretch marks on the skin. Diabetes may be made worse and some previously non-diabetic people may become temporarily diabetic. Some even more serious side-effects are premature arrest of growth in children, cataracts and fractures, and other abnormalities of bone. A patient who stops taking large doses of steroids is in many ways similar to a sufferer from Addion's disease in that stressful events such as having an operation may be complicated by collapse unless steroids are given during the period of stress. This vulnerable period following a prolonged course of high-dose steroids lasts about a year after stopping treatment.

Despite these problems, the steroids can be life-saving in severe asthma, acute allergic disease, and a number of diseases due to altered immunity.

There are techniques to reduce or abolish toxicity by giving the drugs for only a short time or applying the steroids locally. Examples of the latter are injecting steroids directly into a joint in arthritis, inhalation into the lung in asthma, and administration as an enema in colitis.

Diabetes

Diabetes (more correctly, diabetes mellitus) is a condition in which there is a persistently raised blood glucose level. Glucose is the main type of sugar found in the body and is involved in many of the chemical reactions carried out by living cells. Diabetes arises because of a lack of insulin, or failure to respond to it.

Insulin is a hormone secreted by the pancreas and is involved in lowering blood glucose. In one form of diabetes, the cells of the pancreas which produce insulin degenerate. This is most common in diabetes which starts in childhood or early adult life. The form of diabetes which typically starts in later life and may also be associated with obesity is due to a failure of the tissues to respond to insulin. In either case, insulin's actions are reduced. Insulin helps glucose to enter cells so that it can provide energy for several chemical processes necessary for life. These include the manufacture of proteins and fats.

The short-term effect of a high level of blood glucose is that it acts on the kidney as a diuretic and thus increases the rate of urine production. As a result the patient frequently has to empty a full bladder and thirst becomes a prominent symptom. In the long-term, the persistently raised blood glucose concentration is responsible for many of the serious complications of diabetes, such as eye disease, kidney damage, and degeneration of nerves. The abnormalities in fat metabolism cause an increase in blood cholesterol which predisposes to heart attacks, strokes, and blockage of arteries in the legs.

It is therefore a matter of great importance in treating diabetes, to aim to control the blood glucose so that it follows the normal, non-diabetic pattern. This consists of a limited and short-lived rise after food, and a fasting level in the range of 3–5.5 mmol/l. To assist the patient, test strips are used which react to a drop of blood by a colour change that corresponds to the blood glucose level. Another aspect of diabetic treatment is lowering of raised blood cholesterol. This is achieved by a diet which is low in animal and dairy fats, and with partial replacement of these fats with fish and vegetable oils.

Thus, diabetic treatment begins with modification of the diet. Many patients in whom the condition has started in adult life can be successfully managed by diet alone. If the patient is overweight, then a modest reduction in total calorie intake may be required. Rapid weight loss from a very low-calorie diet is dangerous and cannot be sustained for more than a few weeks. Exercise not only assists in weight loss, but it may modify the appetite to be more appropriate to the body's energy needs; it also has an antidiabetic effect by increasing the response of the tissues to insulin.

Yet another feature of the diet in diabetes is to avoid a rapid flow of glucose from food into the blood. Because of this all food containing sugar has to be avoided. Common sugar is sucrose, which is broken down to glucose and fructose in the intestine. Carbohydrate in the form of starch is allowed, because this is only slowly converted into glucose and therefore does not cause a great and rapid increase in blood glucose concentration. Permitted starchy foods include bread, rice, pasta, and potatoes.

Fibre is valuable in the diabetic diet because this indigestible material affects how other foodstuffs are absorbed into the blood. The fibre found in the bean family, guar gum, and pectin slows down glucose absorption into the blood and probably also impairs absorption of fat.

Because total fat in the diet is reduced, there must be a relative increase in other components. The current view is that the proportion of starchy carbohydrates should be increased but not the proportion of protein. There is no case against protein as such, but animal protein cannot be separated from animal fat. Even if all the fat is removed from a piece of meat, the lean parts contain animal fat within the meat fibres themselves.

Some of the older, plumper diabetic patients cannot be successfully treated with diet alone and therefore drugs are used in addition to the diet. These drugs are given by mouth. The majority such as chlorpropamide act by increasing the amount of insulin secreted by the pancreas. Too large a dose can cause an excessive lowering of blood glucose. This can result in feelings of giddiness, sweating, palpitations, tremor, drunken behaviour, or coma. Death can result from very low blood glucose concentrations. Another problem with these drugs is that appetite may be stimulated. This is a particular hardship for patients on a weight-reducing diet. An alternative oral drug is metformin which reduces appetite and acts by slowing the absorption of glucose from the intestine.

Chlorpropamide is given once a day, because its action persists for over 24 hours. Glipizide has a shorter action and is usually taken twice daily. Metformin has a very short action and is therefore given before each meal, taking a larger dose before the biggest meal.

If the diabetes is due to an severe lack of or absence of insulin, then insulin has to be given. Because this substance is destroyed in the stomach and intestine, it has to be injected. The normal secretion of insulin from the pancreas is low during the night, but shows rapid fluctuations during the day—rising on eating and falling during exercise. The current regimens of injecting insulin only once or twice (or less commonly three times) each day are attempts to put maximum amounts of insulin in the blood during and after each meal. The pattern produced only roughly corresponds to normal situations. The result is that excessive swings of blood glucose develop—with over-high levels during or immediately after a meal, because of slower absorption of injected insulin than secretion of insulin from the pancreas. There is a tendency for excessively low levels of blood glucose in the early part of the night because the absorption of insulin from the site of injection takes several hours and no food is taken at this time to balance it. Also, exercise causes a more rapid absorption of injected insulin, which is the opposite to the natural phenomenon. The more carefully the patient can balance insulin injections, spacing and size of meals, and exercise, the better the control of blood glucose levels.

Insulin can cause allergic reactions of several types. Frequently local redness and shrinking of fat at the site of injection is a form of allergy. Even more troublesome is the need to inject increasing amounts of insulin because the patient has developed antibodies in the blood which partially inactivate injected insulin.

Insulin has for many years been prepared from the pancreas of cattle and pigs. The structure of pig insulin is more like human insulin, and this presumably explains why pig insulin causes fewer allergic reactions than cattle insulin does. In recent years, human insulin has become available. This is not obtained from the human pancreas but is obtained in the laboratory. Human insulin causes a very low incidence of allergic responses.

Sex hormones

The sex hormones are made in the ovary, testis, and adrenal cortex. The principal sex hormones secreted by the ovary are oestrogens and progesterone. The main hormone secreted by the testis is testosterone. (See Chapter 9.)

11

The bones, joints, and rheumatism

The bones form the scaffolding which supports the body, while mobility depends on the joints between, and the muscles acting on the skeleton. Of these three tissues, serious muscle diseases are rare, although muscle pains are common. Apart from osteoporosis, diseases of the skeleton are also uncommon; by contrast, diseases of the joints and, in particular the several kinds of arthritis are among the most frequent, chronic, and distressing afflictions of humankind. It should be mentioned incidentally that since the muscles and joints act in concert, muscle and joint pain frequently cannot be distinguished from one another and either can give rise to what is commonly called rheumatism. The latter is not a single disorder but a catch-all term for virtually any disease causing pain on movement, but because of their importance rheumatic conditions will be discussed first.

Rheumatoid arthritis

Rheumatoid arthritis is one of the most common disabling diseases; it causes chronic and often very severe joint pain and can lead to permanent crippling. The cause of the inflammation and consequent damage to the joints is unknown but is regarded as being auto-immune in nature. The disease more commonly affects women, often in their 30s or 40s, and overall may attack 2 per cent of the population.

Small joints such as those of the hands are particularly involved,

145

usually symmetrically. The onset can be insidious or acute with swelling of and pain in the joints, together with wasting of the muscles on either side: fever and anaemia are often associated. In the worst cases, the joints can ultimately become severely damaged, distorted, and almost immobile.

For those who do not suffer from them, it is hard to appreciate the misery that rheumatic diseases cause to millions of people. However, because of their undramatic nature and the fact that they usually become most severe later in life they do not evoke the sympathy or concern that diseases of childhood and infancy, for example, typically evoke. Nevertheless the pharmaceutical industry is aware of the scale of the problem and continues to find new drugs to try to deal with it.

Anti-inflammatory analgesics

Treatment of rheumatoid arthritis is mainly with anti-inflammatory analgesics, of which aspirin is the prototype and still in many ways one of the most satisfactory. The large and varied group of non-steroidal anti-inflammatory drugs (NSAIDs), act by inhibiting production of prostaglandins—the main mediators of inflammation and pain. Nevertheless, prostaglandins also have other beneficial roles and in particular help to protect the stomach lining against the natural acid and digestive secretions. Gastric irritation is therefore one of the most common toxic effects of drugs which inhibit prostaglandin production.

Aspirin was introduced as long ago as 1899 and has not been displaced. Its main effects are suppression of inflammation and consequent pain, and also of fever (antipyretic action) probably by depressing prostaglandin production within the brain. Its toxic effects, in addition to gastric irritation, include interference with the control of bleeding by reducing the stickiness of platelets and preventing them from forming an effective plug. Though this can occasionally prolong bleeding after wounding, it can also be used beneficially to prevent abnormal clotting of blood within the circulation, and possibly even to reduce the risk of heart attacks.

Allergy to aspirin is considerably less common than many believe. It can certainly aggravate asthma in some people but, paradoxically, can improve it in a few others. Overdose is indicated by ringing in the ears (tinnitus) and can impair hearing. In children under 12 years of age, particularly if they are feverish as a result of a viral infection such as chicken-pox, aspirin can very rarely precipitate Reye's syndrome (a dangerous combination of brain and liver damage) and should not be prescribed for this age group unless there is a strong medical reason for doing so.

For rheumatic disease, aspirin needs to be taken regularly and in relatively large doses, namely three or four tablets (300 mg each) every four hours to a total of a dozen or occasionally more tablets a day. To reduce gastric upsets, aspirin should be taken in soluble or dispersible form (preferably dissolved first) with plenty of fluid or after food. In this way even large doses of aspirin can be tolerated. The chief disadvantages are the frequency with which the tablets have to be taken and the loss of traces (usually) of blood from the stomach lining. Over long periods such blood loss can cause anaemia or aggravate the anaemia often associated with rheumatoid arthritis. A few individuals are highly susceptible to this effect so that a few tablets of aspirin can precipitate severe gastric bleeding.

Special preparations of aspirin and of other salicylates, are benorylate (Benoral) an aspirin–paracetamol ester, choline magnesium trisalicylate (Trilisate), (Disalcid), and sodium salicylate. The last, however, is almost obsolete.

It is very difficult to compare objectively the efficacy of other non-steroidal anti-inflammatory drugs. There is also considerable variation in individual susceptibility to the side-effects, particularly gastric irritation from these drugs. As a generalization, the newer (non-aspirin) analgesics are approximately as effective as aspirin but usually better tolerated and may also have the advantage of being longer acting. Since their mode of action is similar to that of aspirin, most of the toxic effects are similar in nature and differ mainly in degree.

Ibuprofen (Brufen) seems to be the safest of this group and is the only one available without prescription (Nurofen). The safety of ibuprofen is probably related to its relatively weak anti-inflammatory activity and it is less effective for severe rheumatic pain such as that of acute gout. Ketoprofen (Orudis) has similar activity to ibuprofen but may be be more prone to cause gastric irritation.

Naproxen (Naprosyn) seems to be one of the most effective of these drugs and significant side-effects are uncommon. Naproxen also has the advantage that it only needs to be given twice daily. Fenoprofen (Fenopron) and flurbiprofen (Froben) are no less or slightly more effective than naproxen but tend to cause more gastric irritation. Azapropazone (Rheumox) is of similar effectiveness to naproxen but more prone to cause rashes, while fenbufen (Lederfen) may be even more prone to cause rashes but has little risk of gastric bleeding. Piroxicam (Feldene) is of similar efficacy but is long-acting and one dose a day is often sufficient. Other drugs similar in their activity, effectiveness, and side-effects include diclofenac (Voltarol), tiaprofenic acid (Surgam), and tolmetin (Tolectin DS).

Indomethacin (Indocid) is one of the older drugs; it has a strong anti-inflammatory effect but frequently causes toxic effects such as headaches, light-headnesses or confusion, and sometimes gastric bleeding. It can also (like the corticosteroids) conceal the onset of infection which can therefore progress dangerously, and can interfere with the action of other drugs such as anticoagulants or antihypertensives (Chapter 5).

Diflunisal (Dolobid) is unusual in that it is an aspirin derivative but is more similar in effect to the non-aspirin analgesics. Benorylate (Benoral) is a compound of aspirin and paracetamol and has properties of both. Mefenamic acid (Ponstan) has only a weak anti-inflammatory action and unlike the others can cause diarrhoea or, rarely, anaemia.

Phenylbutazone (Butazolidine) is a potent anti-inflammatory drug which is effective for severe rheumatic conditions but can cause such severe toxic effects that it is now only available for specialist use. Its chief danger (apart from those in common with other anti-inflammatory drugs) is damage to the bone marrow. This is not always controllable by stopping the drug and fatal aplastic anaemia can follow. Over the years, this has caused hundreds of deaths. Phenylbutazone can also cause rashes, retention of water (with consequent risks of worsening hypertension or heart failure), and sometimes liver damage. Phenylbutazone can also interfere with the action of several other drugs.

Risks and benefits of anti-inflammatory analgesics

The problems caused by phenylbutazone illustrates the dangers associated with some anti-inflammatory drugs. Such drugs are so much needed as a result of the great numbers of patients suffering from depressingly severe and persistent pain or disablement from rheumatic diseases, that there are constant efforts to produce drugs which are not merely more effective but also less prone to cause toxic effects. The result is that there is a bewildering choice of these drugs and it has to be accepted that, for no clear reason, people vary widely in their response to them. An analgesic which works wonders for one person may merely make someone else feel ill.

As discussed earlier, the toxic effects of this group of drugs are largely inherent in their mode of action. Nevertheless different anti-inflammatory drugs vary widely in their side effects: for example, benoxyprofen (Opren: withdrawn because of toxicity) is an effective hair restorer, while, in the newborn, indomethacin can prevent a heart defect known as patent ductus arteriosus, by accelerating its closure.

Other antirheumatic drugs

Anti-inflammatory drugs, though immensely widely used and valuable for the treatment of rheumatoid arthritis, do not control the underlying disease. Some drugs appear, to a greater or lesser degree to do this and include gold salts, penicillamine, chloroquine, and immunosuppressive drugs, but their toxic effects can sometimes be very severe. They are therefore only used when anti-inflammatory analgesics are inadequate to prevent uncontrollable pain and disability. These powerful drugs may also lessen other effects of rheumatoid arthritis on organs other than the joints. Unlike the anti-inflammatory drugs they take several weeks to achieve their effect.

A gold salt in the form of sodium aurothiomalate (Myocrisin) is given by deep injection, initially in small doses to test how it is tolerated and, if it appears safe to continue, at weekly or longer intervals according to the response. If this is good and the drug is well-tolerated it needs to be given indefinitely, as cessation of treatment can precipitate a severe relapse. Toxic effects which preclude continued use of the drug include rashes and severe ulceration of the mouth, while depression of the bone marrow or renal damage can be fatal.

Penicillamine (Distamine; Pendramine) has a similar action to gold salts but toxic effects tend to be less frequent and severe. Nevertheless it can sometimes cause severe allergic reactions as well as nausea, disturbance of taste and loss of appetite, widespread ulceration of the mouth, or damage to the kidneys or bone marrow, among other effects.

Chloroquine is another example of the variety of actions (as opposed to toxic effects) of some drugs. It is primarily an antimalarial drug but is also effective for rheumatoid arthritis and the related disease lupus erythematosus. A major disadvantage of chloroquine is that it can cause irreversible damage to sight if not used with great care, as well as a variety of other side-effects. Nevertheless chloroquine, like other potent anti-rheumatic drugs, is of value in selected cases and toxic effects are by no means inevitable.

Immunosuppressive drugs are used in a variety of auto-immune diseases to depress or damage the body's cells which appear to be causing the disease. Since these cells also form the main line of defence against infection, the activities of these drugs are by no means wholly desirable, especially as they are not particularly selective in their action and can harm other organs such as the bone marrow.

Unfortunately, also, flare-up of the disease can follow cessation of treatment, particularly when corticosteroids, which were once hailed as a panacea for rheumatoid arthritis and many other diseases, are used. For this reason, as well as their many other adverse effects, corticosteroids are no longer favoured for the treatment of this disease. Instead, azathioprine or sometimes, chlorambucil or cyclophosphamide may be given but only by specialists.

Corticosteroids, such as prednisolone, lessen the severity of inflammation of rheumatoid arthritis. Though they may be effective for this purpose in otherwise intractable cases, they do not affect the progress of the disease. Another disadvantage of long-term administration of corticosteroids is weakening of the bones (osteoporosis) and this may lead to fractures or collapse of the spine.

Osteoarthritis

Osteoarthritis is traditionally regarded as being mainly a wear-and-tear disease of overloaded joints, but the causes are far more complex. The joints lose their normal lubricant lining, the cushioning joint surface is destroyed, and the bone ends become exposed and distorted. Joints which have little stress put on them can also be affected and the common observation that osteoarthritis causes disabling hip disease in overweight elderly people results from the fact that the prolonged excessive stress accelerates the damage to the joints and causes them to be painful.

Little can as yet be done to control the disorder which makes joints vulnerable to these changes but the important measures are to lessen the load on affected joints by losing weight—always more easily said than done, particularly as age advances but essential nevertheless— and to increase the strength of the muscles supporting them. Swimming is an ideal form of exercise since the load is taken off the joints.

Such measures may not be possible for the elderly and in many the disease progresses inexorably. Ultimately, it is often necessary to have artificial replacement for hips to regain mobility and relieve pain. However, there is increasing evidence that anti-inflammatory drugs are beneficial for something more than their analgesic action.

Gout

Gout is usually an inborn disorder, particularly of men and leading to excessive levels in the blood of uric acid—a breakdown product of body substances known as purines. Gout is not caused by eating or drinking too much, but over-indulgence can precipitate attacks in those who are predisposed. The main effect is to cause deposition of urate crystals in joints, notably in the big toe, with resulting irritation, inflammation and, in acute attacks, intense pain. Urates can also be deposited in sites such as the ear to form hard white masses of crystals (tophi) which may extrude through the skin.

The management of gout illustrates some of the complexities which sometimes underlie drug treatment in general. Short-term (acute) attacks can be treated as for other rheumatic complaints but the long-term control of the complaint with drugs which modify the disease process is quite different. These drugs used in long-term control can at first, precipitate or worsen acute attacks.

Acute gout, which because of the severity of the pain obviously causes the most concern, is treated with anti-inflammatory analgesics such as naproxen or indomethacin, but not aspirin.

Colchicine is not an analgesic, but has a complex action which makes it specifically effective in relieving the pain only of acute gout. Unfortunately gastro-intestinal and other side-effects can be unpleasant and the drug cannot be tolerated by some.

In the long-term, gout can be controlled by increasing the excretion of uric acid by the kidneys using uricosuric drugs such as probenecid (Benemid) or sulphinpyrazone (Anturan). This a logical way to treat gout but does not relieve acute attacks. Both probenecid and sulphinpyrazone can cause rashes and gastro-intestinal upsets and their action is antagonized by aspirin.

Gout can also be treated with drugs which reduce the formation of uric acid, such as allopurinol (Zyloric). Like uricosuric drugs, allopurinol is effective only in the long-term control of gout and must not be given in an acute attack. It can sometimes cause rashes or more severe allergic reactions, or occasionally severe ulceration of the mouth.

Other forms of arthritis

There are so many kinds of arthritis that it is possible only to consider the main types here. In the past, destructive suppurative infections were once common but should now be manageable with antibiotics and are rarely seen.

Ankylosing spondylitis mainly affects men and involves the spine. Painful inflammation of the spinal joints (spondylitis) leads to fusion of the individual bones (ankylosis) and rigid fixation of the spine often in a flexed position. In the active stages of the disease, the pain has been known to lead to suicide. In severe cases, the chin may ultimately be forced down on to the chest, causing breathing difficulties. Anti-inflammatory analgesics are the main line of treatment for ankylosing spondylitis which is now the only condition for which phenylbutazone may be used, under specialist supervision in hospital. Occasionally, in otherwise unresponsive cases of unusually active disease, corticosteroids may be needed but only for brief periods.

There are several forms of *juvenile arthritis* which affect and can disable children. One of these appears to be a form of rheumatoid arthritis, while another resembles ankylosing spondylitis. Yet another kind of arthritis is associated with the common skin disease, *psoriasis*. Most of these different types of arthritis have to be managed in a generally similar way to rheumatoid arthritis.

Several viruses, such as the hepatitis B and glandular fever (infectious mononucleosis) viruses, can cause arthritis, but this is usually transient and there is no specific treatment available. The arthritis of rheumatic fever after streptococcal throat infection, has virtually disappeared from Britain. In any case, though very distressing, it was reversible in contrast with the possibility of permanent damage to the heart.

Polymyalgia rheumatica

This relatively common disease predominantly attacks those over middle age and causes widespread rheumatism-like pain in the shoulder or hip girdle regions. The pain, however, comes from the muscles (myalgia) due to inflammation in their blood vessels. The same disease but affecting the head can cause severe headache and can lead to blindness as a result of blockage of the arteries to the eyes. This form of the disease, known as cranial arteritis, may be associated with polymyalgia rheumatica or develop on its own.

Polymyalgia rheumatica is one of the few rheumatic conditions for which corticosteroids, such as prednisolone, must be given because of the risk of damage to sight. They must be given until blood tests show that all signs of inflammation have subsided.

Back pain

There are numerous possible causes of back pain, ranging from poor posture at work, injuries or ankylosing spondylitis, to so-called fibrositis (fibromyalgia), or a deposit of cancer in the spine. There is obviously therefore no simple, single line of treatment for such a diversity of conditions, especially as in many cases the causes are not yet fully understood.

One well-known cause of back pain is the extrusion of the soft material from between the bones of the back (commonly called a slipped disc) into the canal. The spinal cord becomes compressed as a consequence and pressure on these nerve fibres commonly then causes intense pain down the leg (sciatica). These disc lesions may gradually heal of their own accord, but can be so severe that it is sometimes necessary to open the spinal canal to remove the protruding matter.

Those with back pain who fail to get relief from orthodox medical treatment may seek relief in manipulation. Undoubtedly this sometimes appears (for no scientifically accepted reason) to be remarkably effective. On other occasions it can do more harm than good—but it is certainly arguable that this may happen with orthodox medicine also.

12

Disorders of skin, eyes, and ears

Skin

One factor which leads some doctors to specialize in dermatology, is that the disease process is visible, and the nature, severity, and extent of the disorder is usually immediately apparent. Unfortunately it is just these factors which increase the misery for the sufferers from skin disorders. To have a rash on a visible part of the body can cause shame and embarrassment and other forms of mental distress, even in the absence of any unpleasant symptoms from the skin. This is a reason why skin complaints account for a high proportion of medical consultations.*

The skin is not an inert covering for the body, but a highly reactive and labile structure. Extensive burns which damage the body surface and impair the normal function of the skin can be fatal. Water, electrolytes, and protein are lost and resistance to infection is weakened. Similarly, in widespread skin disease the patient's life may be endangered because of loss of normal skin function.

Eczema and dermatitis

Dermatitis literally means inflammation of the skin, but in most cases the term is used as a synonym for eczema. The lesions are red, scaly, and itchy.

*Doctors (who should surely know better) frequently write and speak of *skin* rashes, despite the fact that rashes, by definition, *are* diseases of the skin.

Infantile eczema (atopic eczema)

This can begin any time from the early weeks of life up to two years. The face and scalp may be involved, but most frequently the front of the elbows and back of the knees are affected. The children often have an allergic predisposition—in particular, concurrent asthma or hay fever.

Contact dermatitis

This follows exposure of the skin to chemical agents. Any substance is capable of provoking the reaction, but the most commonly implicated are epoxy resin glues, nickel (in rings, earrings, and necklaces), detergents, bleaches, primula plants, and hair dyes.

Varicose eczema

This can develop on the ankle when the leg is swollen due to varicose veins.

   The treatment of eczema is first to remove the cause when this is possible. Thus, in contact dermatitis, it is essential to try to identify the substance which has precipitated the reaction.
   Varicose ulcers and varicose eczema heal if the leg swelling can be eliminated by improving the functioning of the veins, bandaging, or raising the leg to allow the fluid to drain from it.
   Local treatment with steroid lotions, creams, and ointments is effective. Hydrocortisone (1%) is safer than stronger steroids such as betamethasone valerate (Betnovate) and triamcinolone acetonide (Adcortyl; Ledercort). Although local applications of steroids are much less hazardous than taking these drugs by mouth, they are not without problems. The skin of small children is thin and enough steroid can be absorbed through it to cause such generalized adverse effects as delayed growth. The local toxic effects of steroids on the skin include thinning, prominence of blood vessels, aggravation of skin infections, and persistent spots when used on the face. Because of the promotion of

infections, there are many skin preparations of corticosteroids which contain antimicrobials.

Watery applications are helpful for very inflamed, weeping eczema. Such preparations are lotions and simple solutions. Old-fashioned remedies may also be helpful—a very dilute solution of *potassium permanganate* can soothe acute eczema. The main problems with this treatment is that over-concentrated permanganate solutions can burn and, at all strengths, the skin and nails (and the vessel used to contain the solution) are stained.

Psoriasis

This is another common skin disease. Despite intense and prolonged research the cause of psoriasis is not known. The lesions consist of a dark salmon-red thickening of the skin with silvery scaling. The condition may fluctuate in intensity—sometimes emotional stress or infections precipitate a flare-up.

The appearance of the skin and the scaling often cause much shame and embarrassment. It is important for the patient and his/her acquaintances to understand that the disease is not infectious.

Exposure to the sun or artificial ultraviolet radiation and swimming should be encouraged as these may improve the lesions. There are also several skin preparations which can help. Dithranol is a dye which can be incorporated into a paste for application to the patches of psoriasis. This paste may irritate normal skin. The ideal time for applying the paste is after a warm bath in a tar preparation. This should be repeated daily. It increases the shedding of the scale from the lesions, which then become paler and shrink.

Preparations of *coal tar*—(such as crude coal tar paste) are less irritating than dithranol and can be used for people whose skin becomes inflamed even when the treatment is applied accurately to the lesions. Also if the skin trouble is widespread and consists of numerous small plaques, then coal tar in an ointment can be spread all over with little risk of causing any adverse effect on the intervening normal skin.

Treatment techniques can be combined. Thus, a daily tar bath and ultraviolet ray treatment can be followed by careful application of dithranol paste to the major skin lesions.

Steroid creams are sometimes used by dermatologists for certain problems in psoriasis, but this treatment should only be prescribed by experts, as serious (and even dangerous) worsening can follow stopping the steroid. Other drugs which are helpful in expert hands but potentially harmful otherwise are razoxane and methotrexate. Substances called psoralens can be given by mouth two hours before long-wave ultraviolet irradiation to enhance the action of the irradiation. Etretinate is another oral drug only to be prescribed by expert dermatologists. It is a derivative of vitamin A and has a remarkable action in removing excess scale and shrinking the patches of psoriasis. Constant hospital supervision is mandatory with this treatment. Many toxic actions can develop, including damage to the unborn child if given to a pregnant woman: several hundred cases of congenital abnormalities have been caused by this drug in the United States.

Fungal infections

The common term for fungal infections of the skin is ringworm. As well as the skin surface, the nails and hair can also be involved.

Local (topical) treatment to the affected area of skin often cures the infection. Suitable creams and ointments include Whitfield's ointment (which contains benzoic acid), imidazoles (clotrimazole, econazole, miconazole, and ketoconazole), salicylic acid (Phytocil paint), and undecenoates (in Monphytol, Mycota, Phytocil powder and Tineafax). The nails and hair do not usually respond well to local treatment and for these, oral medication is necessary. An older drug for such resistant conditions is griseofulvin. When it is used to treat fungal infection of the nails, treatment may have to be continued for a year. Miconazole and ketoconazole taken orally are probably more effective and more rapidly acting than griseofulvin when used for such problem infections by fungi. However, ketoconazole can occasionally cause liver damage and therefore should only be taken if the severity of the infection merits exposing the sufferer to the risk of drug toxicity and if facilities are available to monitor the liver function (by blood tests) during treatment.

A common infection, thrush, in and at the angles of the mouth is due to the fungus *Candida albicans*. Mouthwashes and other local applications of nystatin are usually effective. Amphotericin is an alternative local treatment for oral thrush. The most common problem with nystatin and amphotericin mouthwashes is their awful taste.

Thrush often infects the vagina and is treated by pessaries or cream of nystatin, clotrimazole (Canesten), econazole (Ecostatin; Gyno-Pevaryl), or amphotericin (Fungilin). If the infection is resistant or recurrent then oral treatment with miconazole tablets is needed as well as local applications. Nystatin taken by mouth may be helpful to rid the gut of the fungus and thus prevent re-infection of the vagina.

Acne

Because acne is common and not a life-threatening condition, its importance is sometimes underestimated. All of us are distressed by any skin abnormality, and adolescents (in whom acne is common) are particularly sensitive about their appearance.

The precise cause of the skin abnormality is not known—but plugs of cells block skin glands, which then form blackheads or become inflamed or infected to become pustules. Apart from these problems due to active disease, it can leave permanent scarring.

Treatment is initially local and its purpose is to clean the skin (to remove excessive grease), to remove the plugs which cause blackheads and inflammation, and to reduce the bacteria in the skin which cause the pustules and other inflamed lesions.

Cleaning the skin with soap or a detergent such as cetrimide is the usual initial advice. If this is done gently then no harm results. Strong antiseptics and agents to cause peeling of the skin can cause irritation unless used with care and moderation. Ointments, creams, and lotions containing such agents include sulphur and resorcinol paste, benzoyl peroxide gel and cream, salicyclic acid and sulphur ointment, sulphur lotion, and scrub solutions containing benzalkonium chloride. A more powerful peeling agent is tretinoin, a derivative of vitamin A which is available as a cream, gel, or lotion.

More resistant acne frequently responds to oral antibiotics such as tetracyclines or erythromycin given in low doses over the course of several months.

In severe acne a new oral treatment is available which, like tretinoin, is a derivative of vitamin A. This is isotretinoin. It is given orally, but has the ability to produce serious toxicity and should only be taken under expert dermatological supervision. It is essential that no pregnant woman takes isotretinoin because of potential harm to the unborn child.

Eyes

Although eye disease is common, it does not often involve primary treatment with drugs. However, there are some eye conditions which can respond to drug treatment. These are inflammatory diseases of the eye and glaucoma.

Conjunctivitis

Conjunctivitis is inflammation of the thin surface covering of the eye. It is commonly caused by micro-organisms. If these are bacteria then the appropriate treatment is usually an antibiotic (such as chloramphenicol or tetracycline) given locally in the form of eye drops or eye ointment. Such drops are rapidly diluted by tears and therefore initially have to be given very frequently (every two hours). Ointment releases the antibiotic for a more prolonged period and such a preparation can be used at night.

Smaller organisms than bacteria include the chlamydia. These can cause conjunctivitis and keratitis—one form of which is trachoma, a tropical infection which can lead to scarring of the eye and blindness. Trachoma responds to tetracycline eye ointment as well as oral sulphonamides or erythromycin.

Viral infections of the surface of the eye are often caused by the herpes simplex virus. This usually responds to acyclovir, idoxuridine, or vidarabine applied locally as eye drops or ointment. Steroids applied to the eye greatly worsen infections by allowing them to spread and ulcerate.

Allergic inflammation

Allergic inflammation of the eye may respond well to steroid eye drops (such as hydrocortisone and dexamethasone), but these can be dangerous because they can cause infections (such as herpetic keratitis) to spread, and they can also aggravate glaucoma by raising the pressure in the eye. Much safer is sodium cromoglycate eye drops (Opticrom) which can ease the itching of the eyes in hay fever sufferers.

Glaucoma

Glaucoma is an abnormally high pressure in the eye. If this is prolonged it can cause blindness. The tendency to glaucoma may be inherited, but acquired diseases such as diabetes or some eye disorders can precipitate it. The usual mechanism is that production of fluid within the eye continues but cannot drain away. The result is high pressure which can damage the retina.

There are surgical and laser treatments available and orally administered drugs may be used, but the most common initial treatment is eye drops. These can provide, in some patients, sufficient control of eye pressure to preserve the sight so long as the treatment is continued.

Many locally applied drugs increase the outflow of fluid in the eye by pulling the tissue of the iris away from the outer rim, where fluid can pass into the blood or lymph. In people taking this type of drug the pupil of the eye becomes smaller than normal. Examples are pilocarpine and physostigmine. The drugs are administered as eye drops every three or four hours. Their most common adverse effects are pain in the brow and blurring of vision. In addition, other substances may be used to potentiate the effects of these first-line drugs. An important example is timolol. This is a beta-blocker (similar to propranolol and atenolol) which is also given as eye drops. It lowers pressure in the eye by reducing the secretion of the fluid within the eye. Even though little timolol is absorbed into the general circulation from the eye, in a few sensitive individuals enough enters the body to cause toxic effects. In particular, sufferers from asthma may find a severe worsening of their chest condition. Over the last year other beta-blockers have been prepared in eye drop form. These are betaxolol, carteolol, and metipranolol. Other drugs as eye drops which may also be effective are adrenaline, dipivefrine, and guanethidine.

Oral drugs may also be given. These decrease the ability of the eye to secrete its internal fluid. The most commonly used is acetazolamide (Diamox). This is usually a most unpleasant drug to take: it often produces indigestion, pins and needles, and loss of energy.

Ears

Many common problems arising in the ear are not at present amenable to drug treatment. However, inflammation and infections of the outer and middle ear are often treated with drugs.

Otitis externa

Otitis externa is inflammation of the outer ear canal. It usually starts as an irritating eczema, but micro-organisms rapidly become established. If the infection becomes a dominant part of the disorder then severe pain can result.

Probably some local irritation of the skin lining of the canal is the starting point of the sequence. For instance, in hot climates increased sweating predisposes to invasion of the skin by fungi. These fungi cause piling up of surface layers of skin and itching which leads to mechanical damage of the skin. In this way moist infected debris builds up in the canal and leads to worse itching and infection. Also with this type of obstruction, soothing and drying lotions cannot get on to the affected areas. A useful technique is to use a 'wick'. This is a strip of gauze impregnated with zinc oxide and ichthammol cream, which has a drying and mild antiseptic action. More modern preparations include steroids (to reduce itching and swelling) with clioquinol (to treat fungal infections) and antibiotic ear drops (such as neomycin and gentamicin).

If the canal is cleaned by mopping and suction then the condition will either resolve spontaneously or at least will allow antimicrobial or drying agents to get into the canal and act on the affected tissue.

When the ear canal is acutely infected and painful, oral (or even injected) antibiotics are used, often with pain-killers such as codeine or paracetamol.

Otitis media

Otitis media is an infection of the middle ear and mainly affects children. Although bacteria are common causative organisms, otitis media can be a complication of a viral infection such as measles. The condition is painful and may have serious consequences such as spread of the infection into the cranial cavity. Alternatively the infection in the middle ear may become chronic and persist for many years.

The majority of attacks of acute otitis media will settle without antibiotic treatment. However, if pus forms in the middle ear then oral or injected antibiotics are needed. If no pus has formed (and thus presumably bacteria are not present) and the condition is due to a viral infection then antibiotics are not effective.

In both types of otitis media simple analgesics, such as paracetamol are given to relieve the pain.

Ear wax

Ear wax is a normal secretion which needs to be removed only if it causes deafness or pressure on the ear-drum. It can be removed by syringing, but first the wax may have to be softened with ear drops. These may be of olive or nut oil, but probably a detergent type of substance— such as docusate sodium—is at least equally effective.

Ménière's disease

Dizziness, deafness, and noise in the ears (tinnitus) may be due to a disturbance of the inner ear called Ménière's disease. Some degree of relief from the giddiness can be provided by betahistine or cinnarizine. However, the disease is usually progressive—and this tendency to worsening is not influenced by drug treatment.

13

Cancer

Cancer is a commonly used term for malignant tumours. A tumour is literally, no more than a swelling but the term is usually used to mean a malignant growth—in medical terminology, a neoplasm. Some neoplasms are benign and readily removed surgically but cancers present a major threat to life. The most common kinds are the carcinomas. These arise from epithelium, the tissue which covers body surfaces, lines the gastro-intestinal tract, and also forms the glands.

Cancer is not a single disease and even cancers of the same type can behave differently in different patients. Most cancers form masses which grow parasitically at the expense of the rest of the body; they are not clearly circumscribed but insidiously invade the surrounding tissues. In addition, malignant tumours form secondary growths (metastases) at distant sites as a result of spread of tumour cells into the lymphatics or bloodstream. Even when apparently completely removed by surgery, recurrence is common and the formation of metastases makes treatment even more difficult.

Quite different kinds of cancer (the leukaemias) affect the white cells of the blood and do not form solid tumours. Though uncommon, they are an important form of cancer in children.

Most solid tumours are diseases of middle age and over, and the most common sites are the lung, breast, prostate, stomach and large intestine. Despite research extending over more than a century, remarkably little is known for certainty about their causes. The link between cigarette smoking and cancer of the lung is, however, clear. There are also associations between smoking and cancer in other sites, such as the stomach and bladder. Evidence is emerging of a role in cancer of faulty diet such as lack of fibre or citrus fruits, or an excess of salt-preserved foods. Other preventive measures of proven value, such as protection against exposure to a variety of chemicals, have been applied for so long in industry as often to be forgotten.

The effects of cancers are very varied. They also depend on the organ in which the tumour first arises, on its rate of growth and spread, and on many other factors. However, many cancers cause no symptoms in the early stages, when treatment is likely to be most effective. Though cancer is notorious as a cause of unbearably severe pain, this is often a late effect.

Cancer of the lung (one of the most common cancers) for example usually causes no symptoms at first but occasionally the growth may be seen by chance in a routine radiograph. Later, typical features of cancer of the lung are cough (sometimes with blood in the sputum), increasing breathlessness, loss of weight, and chest pain or complications such as pneumonia. However (illustrating the variety of possible effects of cancer), the first manifestation may be changes in the finger tips and nails (clubbing) or muscle weakness, as a result of production of hormone-like substances by the tumour. Alternatively a bone may unexpectedly become painful or even break as a result of destruction by a secondary deposit. Some of these complications are not unique to cancer of the lung.

Anticancer treatment

Historically the first type of treatment of cancer was by surgery (often mutilating) in the attempt to remove all traces of the tumour. Surgery is still the most important mode of treatment of the common cancers.

Radiotherapy (X-rays or other types of radiation) is another method of treatment which is sometimes strikingly effective for some skin tumours or lymphomas. It is frequently also used in conjunction with surgery.

More recently, drugs have been introduced which, in some kinds of cancer are curative without any other form of treatment. Cancers which can be treated successfully with drugs include acute lymphatic leukaemia and some other tumours in children, Hodgkin's disease, testicular tumours, African (Burkitt's) lymphoma, and choriocarcinoma. Total cure of many of the common tumours is possible, particularly if they are discovered at an early stage and before they have become too widespread. In other cases, modern treatment can provide useful prolongation of life, often without significant disability. The limitations of anticancer drug treatment are that not all patients respond in the desired fashion and the price that has to be paid is that toxic effects may be intolerable.

There is also no single effective treatment and each tumour seems to require the administration of a complex mixture of drugs to get the best results. However, the most common cancers—the carcinomas—are best treated by operation initially.

Anticancer drugs (cytotoxic chemotherapy)

Because of the toxic effects of anticancer drugs, they should only be given for tumours which are known to be susceptible. Nevertheless, chemotherapy may also be helpful in palliating incurable tumours.

The weakness of currently available anticancer drugs is that they are relatively unselective in their action. They are mostly cell poisons which take advantage of the fact that tumour cells are multiplying faster than most normal cells. The consequence is that rapidly dividing normal cells such as those of the gastro-intestinal tract, bone marrow, or hair follicles tend also to be damaged. Bone marrow depression is important because it can lead to severe infections as a result of stopping white cell production. Anaemia and bleeding tendencies are other possible complications.

The toxic effects of anticancer drugs, particularly prolonged nausea and vomiting, can be distressingly severe and require expert management. Indicative of their poisonous nature is the fact that early anticancer drugs were derived from the chemical weapon mustard gas.

Because of the great variety of anticancer drugs only a few examples of the main groups will be described. Moreover, since combinations of these drugs are usually given to obtain the greatest benefit and toxic effects can even vary with the route of administration, these descriptions of the effects of individual drugs can only form a general guide.

Alkylating agents

Alkylating agents such as cyclophosphamide, chlorambucil, and busulphan are mainly used for tumours of the white cells such as leukaemias, lymphomas, and Hodgkin's disease. Cyclophosphamide can cause haemorrhagic cystitis (bladder pain and bleeding into the urine), while chlorambucil and busulphan are more prone to damage the bone marrow. Cyclophosphamide can cause total loss of hair (as can several other anticancer drugs) and though there is no way of treating this, the hair eventually grows again after completion of treatment and is sometimes then (as a small consolation) attractively curly and fairer than before.

Cisplatin has an alkylating action and is effective against some solid tumours particularly of the ovaries and testes. However, it causes unbearable nausea and vomiting, can damage the kidneys, depress the marrow, and impair hearing.

Cytotoxic antibiotics

Cytotoxic antibiotics were originally intended to be used against microbes, but found to be so toxic as to have an antitumour effect resembling that of irradiation. One of the most useful of these is doxorubicin which is used for leukaemias, lymphomas, and various solid tumours. It commonly causes nausea and vomiting, ulceration of the mouth, depression of the bone marrow and can damage the heart. Bleomycin has similar uses but, in addition to ulceration of the mouth, can cause a variety of skin disorders and allergic reactions. Its principal complication is progressive fibrosis of the lungs causing increasing difficulty in breathing.

Antimetabolites

Antimetabolites block processes within the cell that lead to cell division. Methotrexate which is used for childhood leukaemias, lymphomas and various solid tumours tends to cause severe ulceration of the mouth, and depress the bone marrow. Cytarabine is even more prone to cause bone marrow depression, while azathioprine so effectively depresses immune responses that it is more often used for such purposes as organ transplantation than for the treatment of cancer.

Vinca alkaloids

Vinca alkaloids are derived from the periwinkle plant, (*Vinca rosea*) which in herbal folklore was believed to have medicinal value. Those with excessive faith in herbal products may care to note that the Vinca alkaloids are highly toxic. Vincristine, in particular, can damage the nervous system causing abdominal bloating and distension, and constipation; there may be paraesthesias (pins and needles) or even muscle weakness. Vinblastine, by contrast, is more prone to suppress bone marrow function.

Hormones, hormone antagonists and other anti-cancer drugs

Cancer of the breast is frequently hormone-dependent. As a result, one treatment was to remove surgically not merely the breast tumour but also all tissues which could produce any oestrogen. This could be called, in surgical terms, an heroic procedure, but it is rather that the patient had to be a heroine to withstand it. Tamoxifen (a hormone antagonist) achieves the same or a better effect by chemically blocking oestrogen receptors in the body. This appears to be a major advance in the treatment of breast cancer, particularly in older women.

Prostatic cancer is androgen (male sex hormone) dependent. Once the primary tumour has been removed, metastases may be controlled by removing the testes or giving the oestrogenic hormone, stilboestrol. Inevitably, this causes feminization, for example growth of breasts (gynaecomastia). Nausea, fluid retention, and thromboses are other toxic effects. Buserelin, which inhibits male hormone release, is one of several alternatives, and has fewer side-effects than stilboestrol.

Interferon

Interferon, originally introduced as an antiviral drug (Chapter 3), has proved highly effective against a rare type of leukaemia. An unpleasant toxic effect of this drug is a prolonged, feverish, flu-like illness.

Other aspects of anticancer treatment

Anti-emetics

Anti-emetics have frequently to be given with anti-cancer drugs to combat nausea and vomiting; these are described in Chapter 6. However, even these drugs may not be entirely effective and patients have been known to refuse to complete anticancer treatment because of severe and distressing nausea. Nabilone, a cannabis-like drug, may be given with anti-emetic drugs to overcome this problem.

Pain control

The idea that one has a potentially lethal disease is difficult enough to face, but that it may be agonizingly painful as well (the traditional idea of cancer) can be too much. Until relatively recently, many doctors may have treated the symptoms of advanced cancer inadequately and, partly because of excessive fear of their adverse effects, may have been parsimonious with drugs to control pain. Moreover, the appreciation has come slowly that the combination of fear with pain or unpleasant effects of treatment, causes severe anxiety and often depression, and that this may aggravate the perception of pain. Cancer pain is frequently, however, controllable, particularly if analgesics are given when pain is anticipated rather than later when it has become established. When given in this way even simple analgesics such as aspirin can be suprisingly effective, but if the worst comes to the worst opiates should be given liberally. It is equally important to maintain morale by open discussion, advice, and reassurance and, if necessary, antidepressant drugs. By pioneering changes such as these, hospices such as St Christopher's have brought about remarkable improvements in cancer care.

The future

All that has been said here may read as a pretty bleak account of the limitations of current anticancer treatment. Though there have been major advances, the toxic effects (particularly prolonged nausea) may make such treatment difficult to bear. On the credit side, cancer treatment is continually improving and existing drugs are being used more effectively in different combinations.

It also seems likely that we are on the threshold of major advances in anticancer treatment in that drugs which are more precisely targeted against tumour cells are in process of development. Specific actions against cancer cells are promised by making use of the immune response both to recognize cancer cells and carry drugs to them alone. These drugs should be considerably more effective and lack the blunderbuss action which is largely responsible for the toxic effects of present drugs.

Ready reference section

Only the more important drugs have been discussed in the text but there are very many alternatives and an even greater number of alternative names for the same drugs. The purpose of this section is to help you to identify the nature of a preparation with an unfamiliar name. For example, if you react adversely to aspirin you can find out if a particular drug preparation contains aspirin.

The reason for the confusing variety of names for drugs is that, in addition to its official (generic) name, a drug will usually have several proprietary (brand) names, given by different manufacturers. (In some cases there may be minor differences in these proprietary formulations, but we have not included such details here.) Unfortunately also, drugs of quite disparate nature can sometimes have confusingly similar names. Examples include *Dyazide*, a diuretic and diazoxide, a powerful drug given by injection for lowering blood pressure. Other examples are *Maalox*, an antacid for indigestion, and *Maxolon*, a powerful antinauseant which acts on the brain. Brand names are distinguished in the list below by being printed in *italics*; in the text they usually appear in brackets after the generic name and always have an initial capital.

In this section it is not feasible to do more than state the general nature and uses for the various drugs. More information about a drug or a closely related one which has similar effects is given in the appropriate chapter. The main omissions from this list are among the almost numberless skin dressings and other preparations, nutritional support preparations, and drugs used in anaesthesia.

All the drugs included here are licensed products (approved by the Committee on Safety of Medicines). Some are considerably more effective or safer than others for a similar application. In other cases, there are several alternative drugs with such minor differences in effect that the choice depends largely on the personal preference of the physician.

Though the main uses for individual drugs are suggested here, this does not necessarily mean that they are completely effective, and in clinical practice they may have little value. For example, drugs which dilate blood vessels are not (contrary to what might be hoped) helpful in cerebral artery disease in the elderly as, among other reasons, they do not relieve physical obstruction of the vessels by atheroma.

Drugs marked CD (Controlled Drug) are prone to addiction and require special prescriptions.

Abidec. A multivitamin preparation.

Accupen. Injection device for insulin.

Accupro (quinapril). For hypertension.

Acebutolol. Beta-blocking drug; many uses including hypertension and angina pectoris.

Acepril (captopril). For hypertension.

Acetazolamide. Diuretic; particularly used for lowering pressure in the eye in glaucoma.

Acetohexamide. Oral drug for lowering blood sugar in diabetes.

Acetomenaphthone. Form of vitamin K used only in vitamin tablets.

Acetoxyl (benzoyl peroxide). Application for acne.

Acetylcysteine. Used to make the sputum more fluid, particularly in cystic fibrosis.

Acezide (captopril with hydrochlorthiazide). For hypertension.

Achromycin (chlortetracycline). Antibacterial agent.

Aci-Jel (acetic acid). Vaginal antiseptic.

Acipomox. For lowering blood lipid levels.

Aclomethasone. Steroid skin preparation.

Acnegel (benzoyl peroxide). Application for acne.

Acrilyse (alteplase). For dissolving coronary thromboses.

Acrivastin. Antihistamine for allergies such as hay fever.

Acrosoxacin. Antibacterial drug for penicillin-resistant gonorrhoea.

Actal (alexitol sodium). Antacid for dyspepsia.

ACTH (corticotrophin). Stimulates cortisone production by the adrenals.

Actidil (triprolidine hydrochloride). Antihistamine for minor allergies.

Actifed (ephedrine preparation). Cough mixture and nasal decongestant.

Actilyse (alteplase). For dissolving coronary thromboses.

Actinac (chloramphenicol and corticosteroid lotion). For acne.

Actinomycin D. Anticancer drug.

Actonorm. Antacid for dyspepsia.

Actraphane. Human insulin for diabetes.

Actrapid. Human insulin for diabetes.

Acupan (nefopam hydrochloride). An analgesic.

Acyclovir. Antiviral drug effective against herpes infections including shingles.

Adalat (nifedipine). Antihypertensive drug.

Adcortyl (triamcinolone acetonide). Corticosteroid preparation.

Adrenaline. Injection for allergic emergencies. Also eye drops for dilating the pupils.

Aerolin (salbutamol inhalation). For asthma.

Aerosporin (polymyxin b sulphate). An antibacterial.

Afrazine (oxymetazoline hydrochloride). Nasal decongestant.

Agarol. Stimulant and lubricant laxative.

Agiolax (senna and ispaghula). Stimulant laxative.

Aglutella. Gluten-free preparations for coeliac disease.

Akineton (biperiden). For Parkinson's disease.

Akrotherm. Application for chilblains.

Albucid (sulphacetamide). Antibacterial eye drops.

Alcobon (flucytosine). Systemic antifungal drug.

Alcopar (bephenium). For hookworm infestation.

Aldactide (spironolactone with hydroflumethiazide). Combined diuretic.

Aldactone (spironolactone). Diuretic for oedema.

Aldomet (methyldopa). For hypertension.

Alexan (cytarabine). Anticancer drug.

Alexitol sodium. Antacid for dyspepsia.

Alfacalcidol. Vitamin D.

Alfentanil (CD). An intravenous opioid.

Algesal. Rubifacient counterirritant.

Algicon. Antacid for dyspepsia.

Alginates. Included in many antacids to protect against reflux oesophagitis.

Algipan. Rubifacient counterirritant.

Algitec (cimetidine with alginate). For peptic ulcers and acid reflux.

Alka-Donna. Antacid with antispasmodic, for dyspepsia.

Alkeran (melphalan). Anticancer drug.

Alkylating agents. Anticancer drugs.

Allbee with C. Multivitamin preparation.

Allegron (nortriptyline). Antidepressant.

Aller-eze (clemastine). Antihistamine for minor allergies.

Allopurinol. Lowers uric acid levels for prophylaxis of gout.

Allyloestrenol. A progesterone used to treat habitual abortion.

Almasilate. Aluminium magnesium silicate compound for dyspepsia.

Almazine (lorazepam). Diazepam-like anxiolytic and hypnotic.

Almevax. Rubella vaccine.

Almodan (amoxycillin). Broad spectrum antibiotic.

Alophen. Stimulant laxative.

Aloral (allopurinol). For prophylaxis of gout.

Aloxiprin. Buffered aspirin.

Alpha tocopheryl acetate. Vitamin E.

Alphaderm (hydrocortisone cream). For inflammatory skin disorders.

Alprazolam. Diazepam-like anxiolytic.

Alprostadil. For emergency treatment of some congenital heart diseases of the newborn.

Alrheumat (ketoprofen). Anti-inflammatory analgesic.

Altacite plus. Antacid for dyspepsia.

Altaplase. For dissolving coronary thromboses.

Alu-Cap (aluminium hydroxide). Antacid for dyspepsia.

Aludrox (aluminium and magnesium compound). Antacid for dyspepsia.

Aluhyde. Antacid with antispasmodic.

Aluline (allopurinol). For prophylaxis of gout.

Aluminium acetate lotion. Application for exudative eczema.

Aluminium hydroxide and glycinate. Antacids for dyspepsia.

Aluminium hydroxide–magnesium carbonate. Compound antacid for dyspepsia.

Alunex (chlorpheniramine maleate). Antihistamine for minor allergies.

Alupent (orciprenaline sulphate). Bronchodilator for asthma.

Alupram (diazepam). Anxiolytic.

Aluzine (frusemide). Potent diuretic for oedema.

Alvercol (alverine). Fibre for irritable bowel syndrome.

Alverine. For irritable bowel syndrome.

Alverine citrate. Antispasmodic for irritable bowel syndrome.

Amantadine. Increases mobility in Parkinson's disease but may also be given for prophylaxis of influenza A.

Ambaxin (bacampicillin hydrochloride). Broad-spectrum penicillin.

Ambilhar (niridazole). For guinea-worm infestation.

Ambutonium bromide. Antispasmodic for dyspepsia.

Amethocaine. Surface anaesthetic.

Amfipen (ampicillin). Broad spectrum antibiotic.

Amikacin. Streptomycin-like antibiotic.

Amikin (amikacin). Streptomycin-like antibiotic.

Amilco (amiloride with hydrochlorothiazide). Combined diuretic for oedema.

Amiloride. Mild, potassium-sparing diuretic.

Aminobenzoic acid lotion. Sunscreening lotion.

Aminoglutethamide. For breast cancer.

Aminoglycosides. The streptomycin group of antibiotics.

Aminophylline. For relief of airways obstruction in asthma.

Amiodarone. Relieves certain disorders of heart rhythm.

Amitriptyline. Tricyclic antidepressive.

Ammonia and ipecacuanha mixture. Expectorant for cough.

Ammonium chloride mixture. Expectorant for cough.

Amoxapine. Antidepressant.

Amoxidin (amoxycillin). Broad-spectrum penicillin.

Amoxil (amoxycillin). Antibiotic.

Amoxycillin. Broad-spectrum oral penicillin.

Amphetamines (CD). Stimulants which suppress appetite (not recommended).

Amphotericin. Antifungal antibiotic.

Ampicillin. Broad-spectrum oral penicillin.

Ampiclox (ampicillin with cloxacillin). Broad-spectrum antibiotic.

Amsacrine. Anticancer drug.

Amsidine (amsacrine). Anticancer drug for leukaemia.

Amylobarbitone (CD). Barbiturate for intractable insomnia, but not usually recommended.

Amytal (CD) (amylobarbitone). See Amylobarbitone.

Anacal. Contains prednisolone for relief of anal irritation.

Anaflex (polynoxylin). For minor skin infections.

Anafranil (clomipramine). Antidepressive drug.

Anapolon (oxymethalone). For aplastic anaemia.

Ancrod. Anticoagulant for vein thromboses.

Androcur (cyproterone acetate). Anti-androgen, for prostatic cancer or male hypersexuality.

Androgens. Male sex hormones. May be used for advanced breast cancer.

Andursil (aluminium and magnesium compound). Antacid for dyspepsia.

Anethaine. Amethocaine, local anaesthetic cream.

Anexate (flumazenil). Antagonist to, and antidote for overdose of, diazepam-like drugs.

Angettes-75. Low dose aspirin to prevent heart attacks.

Angilol (propranolol hydrochloride). For hypertension, angina, and many other purposes.

Angiotensin-converting enzyme inhibitors. A group of antihypertensive drugs.

Anistreplase. For dissolving coronary thromboses.

Anodesyn. For soothing anal irritation.

Anorectics. Drugs to reduce appetite.

Anquil (benperidol). Antipsychotic for deviant and antisocial behaviour.

Antabuse (Disulfiram). For management of alcohol dependence.

Antazoline. Antihistamine eye drops for allergic conjunctivitis.

Antepar (piperazine). For worm infestations.

Antepsin (sucralfate). For control of peptic ulceration.

Anthisan. As for *Anthical*.

Anthranol (dithranol preparations). For psoriasis.

Anthraquinones. Stimulant purgatives.

Anti-androgens. For prostatic cancer or for male hypersexuality.

Antihelmintics. Drugs for worm infestations.

Antipressan (atenolol). For high blood pressure.

Antoin (aspirin and codeine-containing analgesic). For pain.

Antraderm. Dithranol preparation for psoriasis.

Anturan (sulphinpyrazone). For long-term control of gout.

Anugesic-HC. For anal irritation.

Anusol and *Anusol-HC.* For anal irritation.

Anxon (ketazolam). Diazepam-like anxiolytic.

Apisate (CD) (diethylpropion hydrochloride). For appetite suppression.

Apresoline (hydralazine hydrochloride). Antihypertensive drug.

Aprinox (bendrofluazide). Diuretic for oedema or hypertension.

Aprotinin. For prevention or treatment of pancreatitis.

Apsac (anistreplase). For dissolving coronary thromboses.

Apsifen (ibuprofen). Anti-inflammatory analgesic.

Apsin VK. Oral penicillin.

Apsolol (propranolol). For hypertension or angina.

Apsolox (oxprenolol). For hypertension or angina.

Aradolene. Counterirritant ointment for fibrositis.

Aramine (metaraminol). For raising blood pressure in shock.

Aredia (disodium pamidronate). For lowering blood calcium and for Paget's and some other bone diseases.

Arelix (piretanide). Diuretic for hypertension.

Arilvax. Yellow fever vaccine.

Arobon (ceratonia). For diarrhoea.

Arpicolin (procyclidine hydrochloride). For Parkinson's disease.

Arpimycin (erythromycin). Antibacterial mixture.

Arret (loperamide hydrochloride). Antidiarrhoeal capsules.

Artane (benzhexol hydrochloride). For Parkinson's disease.

Artracin (indomethacin). Anti-inflammatory analgesic.

Arvin (ancrod). Anticoagulant for prevention of thromboses.

Arythmol (propafenone). For abnormal heart rhythms.

Asacol (mesalazine). For maintenance of remission in ulcerative colitis.

Ascabiol (benzyl benzoate). For scabies and louse infestations.

Ascorbic acid. Vitamin C.

Asendis (amoxapine). Antidepressant.

Asilone. Antacid for dyspepsia.

Asmaven (salbutamol). Bronchodilator for asthma.

Asparaginase. Anticancer drug.

Aspav. Aspirin-containing analgesic.

Aspellin. Rubifacient counterirritant.

Aspirin. Anti-inflammatory analgesic.

Astemizole. Antihistamine; for minor allergies.

AT 10. Dihydrotachysterol (vitamin D).

Atarax (hydroxyzine hydrochloride). Anxiolytic.

Atenolol. Antihypertensive and anti-anginal beta-blocker drug.

Atensine (diazepam). Anxiolytic and hypnotic.

Ativan (lorazepam). Diazepam-like anxiolytic.

Atromid-S (clofibrate). For lowering blood lipids.

Atropine. Relieves gut muscle spasm and slows heart rate. Dilates the pupils. Present in many different preparations.

Atrovent (ipratropium bromide). Bronchodilator for asthma and bronchitis.

Attenuvax. Measles vaccine.

Augmentin (amoxycillin and clavulanic acid). Penicillinase-resistant broad-spectrum oral penicillin.

Auranofin. Oral gold preparation for rheumatoid arthritis.

Aureomycin (chlortetracycline hydrochloride). Antibacterial agent.

Aurothiomalate. Gold injection for rheumatoid arthritis.

Aventyl (nortriptyline). Tricylic antidepressive.

Avloclor (chloroquine). For prophylaxis and treatment of malaria.

Avomine (promethazine). Antihistamine for minor allergies or nausea.

Axid (nizatidine). For control of peptic ulcer.

Azactam (aztreonam). Reserve antibiotic.

Azamune (azathioprine). Immunosuppressive drug.

Azapropazone. Non-steroidal anti-inflammatory analgesic.

Azatadine. Antihistamine; for minor allergies.

Azathioprine. Cytotoxic immunosuppressive drug mainly used for auto-immune disease or as adjunctive drug for organ transplantation.

Azidothymidine (zidovudine). Antiviral drug for AIDS.

Azlocillin. Special-purpose penicillin.

AZT (azidothymidine). For treatment of AIDS.

Aztreonam. Reserve antibiotic.

Bacampicillin. Broad-spectrum oral penicillin.

Bacillus Calmette-Guerin. For prophylaxis against tuberculosis.

Baclofen. For skeletal muscle spasm.

Bacticlens (chlorhexidine). For skin disinfection.

Bactrian (cetrimide). For skin disinfection.

Bactrim (co-trimoxazole). Broad-spectrum antibacterial agent.

Bactroban (mupirocin). For bacterial skin infections.

Balmosa. Rubifacient counterirritant.

Banocide (diethylcarbamazine). Filaricide.

Baratol (indoramin). Anti-hypertensive drug.

Barquinol HC (hydrocortisone acetate and clioquinol). Anti-inflammatory skin cream.

Baxan (cefadroxil). Broad-spectrum antibiotic.

Baycaron (mefruside). Diuretic for oedema.

Bayolin. Rubifacient counterirritant.

Baypen (mezlocillin). A reserve penicillin.

BC 500. Multivitamin tablets.

BCG. Bacillus Calmette-Guerin. For prophylaxis of tuberculosis.

Becloforte (beclomethasone dipropionate inhalation). For asthma.

Beclomethasone dipropionate. A corticosteroid for asthma or hay fever.

Beconase. Beclomethasone dipropionate nasal aerosol.

Becosym. Vitamin B mixture.

Becotide (beclomethasone dipropionate). Corticosteroid inhalation for asthma.

Bedranol (propanolol). For high blood pressure and angina.

Belladonna. Antispasmodic, present in many antacids.

Bellocarb. Antacid with belladonna (antispasmodic) for dyspepsia.

Benadon (pyridoxine, vitamin B_6). For isoniazid-induced peripheral neuritis or deficiency states.

Bendrofluazide. Diuretic for hypertension or oedema.

Benemid (probenecid). For control of gout or to delay excretion of penicillin.

Benerva Compound (thiamine; vitamin B$_1$). For deficiency states and neurological changes, especially those related to chronic alcoholism.

Benethamine penicillin. Long-acting penicillin.

Benoral (benorylate). Analgesic.

Benorylate. Analgesic which releases aspirin and paracetamol in the body.

Benoxyl preparations (benzoyl peroxide applications). For acne.

Benperidol. Antipsychotic drug. May be given for deviant behaviour.

Benserazide. Increases the action of levodopa (allowing dose reduction) in Parkinson's disease.

Bentex (benzhexol hydrochloride). For Parkinson's disease.

Benylin. Antihistamine-containing expectorant or sedative cough mixtures, or nasal decongestant in different formulations.

Benzagel (benzoyl peroxide gel). For acne.

Benzalkonium chloride. Skin disinfectant.

Benzathine penicillin. Long-acting oral penicillin.

Benzhexol. For mild Parkinson's disease or with levodopa for more severe disease.

Benzocaine. Surface anaesthetic for the mouth or throat.

Benzodiazepines. Includes diazepam (*Valium*) and many related drugs for anxiety and insomnia.

Benzoic acid ointment. Antiseptic ointment for ringworm.

Benzoyl peroxide. Antiseptic present in skin applications for acne and psoriasis.

Benzthiazide. Diuretic for lowering blood pressure and oedema.

Benztropine. Anti-Parkinsonian drug similar to benzhexol but sedating.

Benzydamine. Rinse or spray for painful oral conditions.

Benzyl benzoate. Parasiticide for scabies.

Benzylpenicillin. Penicillin for infection.

Bephenium. Anthelmintic for hookworm.

Berkatens (verapamil hydrochloride). For disorders of heart rhythm, angina, or hypertension.

Berkmycen (oxytetracycline). Broad-spectrum antibiotic.

Berkolol (propranolol hydrochloride). Beta-blocker for hypertension and angina.

Berkozide (bendrofluazide). Diuretic for hypertension or oedema.

Berotec (fenoterol hydrobromide). Bronchodilator inhalation for asthma and bronchitis.

Beta-adalat (atenolol and nifedipine). For high blood pressure and angina.

Beta-Cardone (sotalol hydrochloride). Beta-blocker for hypertension or angina.

Betadine (povidone–iodine). Preparations for skin disinfection.

Betadren (pindolol). Beta-blocker for hypertension or angina.

Betahistine. For control of Ménière's disease (vertigo and hearing disturbances).

Betaloc (metroprolol tartrate). Beta-blocker for hypertension or angina.

Betamethasone. Anti-inflammatory and anti-allergic corticosteroid.

Betaxolol. Antihypertensive beta-blocking drug.

Bethanechol. Stimulant laxative and may relieve retention of urine.

Bethanidine. Antihypertensive drug.

Betim (timolol maleate). Beta-blocker for hypertension or angina.

Betnelan (betamethasone). Anti-inflammatory corticosteroid.

Betnesol (betamethasone sodium phosphate). Solution for inflammation of eye, nose, and external ear canal.

Betnovate (betamethasone valerate). For skin or mucosal inflammation.

Betoptic (betaxolol hydrochloride). Beta-blocker eye drops for glaucoma.

Bextasol (betamethasone valerate). Inhalant for asthma.

Bezafibrate. Lowers blood cholesterol levels.

Bezalip (bezafibrate). For high blood cholesterol.

Bicillin. Long-acting (procaine and benzyl) penicillins.

BiCNU (carmustine). Anticancer drug.

Biguanides. Given by mouth to lower blood sugar in mild diabetes.

Bilarcil (metriphonate). For schistosomiasis infestation.

BiNovum (ethinyloestradiol and norethisterone). Oral contraceptive.

Biogastrone (carbenoxolone sodium). For peptic ulceration.

Biophylline (theophylline). Syrup for asthma and bronchitis.

Bioplex (carbenoxolone sodium). For application in the mouth.

Bioral gel. As for *Bioplex*.

Biotin. Part of the vitamin B complex.

Biperiden. Benzhexol-like drug for Parkinson's disease.

Bisacodyl. Stimulant laxative.

Bismodyn (bismuth subgallate). Anal ointment or suppository.

Bismuth chelate. For treatment of peptic ulceration.

Bisoprolol. For high blood pressure and angina.

BJ6 eye drops. Hypromellose preparation for dry eyes.

Bleomycin. Anticancer antibiotic.

Blocadren (timolol maleate). Beta-blocker for hypertension or angina.

Bocasan (sodium perborate). Antiseptic mouth rinse.

Bolvidon (mianserin hydrochloride). Antidepressant.

Bonjela (choline salicylate application). For mild oral ulcers.

Bradilan (nicofuranose). For poor peripheral circulation.

Bradosol (domiphen bromide). Antiseptic lozenges.

Brasivol (aluminium oxide). Cleansing preparation for acne.

Bretylate (bretylium tosylate). For disorders of heart rhythm.

Bretylium. For disordered heart rhythms.

Brevinor (oestrogen and progestogen). Combined oral contraceptive.

Bricanyl (terbutaline sulphate). Bronchodilator preparations for asthma.

Brietal Sodium (methohexitone sodium). Intravenous barbiturate anaesthetic.

Britiazim (diltiazem). For angina pectoris.

Brocadopa (levodopa). For Parkinson's disease.

Broflex (benzhexol). For Parkinson's disease.

Brolene (propamidine isethionate). Anti-infective eye drops.

Bromazepam. Diazepam-like drug for anxiety.

Bromocriptine. Alternative treatment for Parkinson's disease but also reserve treatment for infertility and excessive lactation.

Brompheniramine. Antihistamine for minor allergies.

Bronchodil (reproterol hydrochloride). Salbutamol-like broncho-dilator for asthma.

Broxil (phenethicillin). An oral penicillin.

Brufen (ibuprofen). Anti-inflammatory analgesic.

Brulidine (dibromopropamidine isethionate). Application for minor skin infections.

Buccastem (prochlorperazine). For nausea.

Budesonide. Corticosteroid inhalation for asthma or nasal allergy.

Bufexamac. Application for mild skin inflammation.

Bumetanide. Powerful diuretic for oedema.

Bupivicaine. Long-acting local anaesthetic.

Buprenorphine. Morphine-like analgesic.

Burinex (bumetanide). Diuretic for hypertension or oedema.

Buscopan (hyoscine butylbromide). Antispasmodic for digestive disorders.

Buserelin. Hormone antagonist for prostatic cancer.

Buspar (buspirone). For anxiety.

Buspirone. For anxiety.

Busulphan. Anticancer alkylating drug.

Butacote, Butazolidin, Butazone (phenylbutazone). Anti-inflammatory analgesics.

Butobarbitone (CD). Barbiturate for intractable insomnia.

Butriptyline. Tricyclic drug for depression.

Butyrophenones. Antipsychotic drugs such as droperidol and haloperidol.

Cafadol (paracetamol and caffeine). Analgesic.

Cafergot (ergotamine tartrate with caffeine). For migrainous attacks.

Caffeine. Weak stimulant; often included in analgesics but without proven value.

Caladryl. Antihistamine cream or lotion; for minor skin allergies.

Calciferol. Vitamin D.

Calcimax. Multivitamin preparation.

Calciparine (heparin calcium). Anticoagulant for prevention of thromboses.

Calcisorb (sodium cellulose phosphate). To reduce calcium absorption.

Calcitare (calcitonin). For Paget's disease of bone.

Calcitonin. Hormone for Paget's disease of bone or to lower blood calcium.

Calcitriol. Vitamin D analogue.

Calcium carbonate. Component of some antacids but unsatisfactory.

Calcium channel blockers. Drugs for treating hypertension and angina, and possibly for preventing migraine.

Calcium folinate and calcium leucovorin. Antidote for methotrexate and other antifolate anticancer drugs.

Calcium gluconate and calcium lactate. Remedies for calcium deficiency.

Calichew. Chewable calcium carbonate for calcium deficiency.

Calpol (paracetamol suspension). For fever and pain in children.

Calsynar (salcatonin). For Paget's disease of bone.

Calthor (ciclacillin). Broad-spectrum penicillin.

CAM. Compound bronchodilator for asthma.

Camcolit (lithium carbonate). For manic–depressive states.

Canesten (clotrimazole). Antifungal applications.

Canesten-HC (clotrimazole with hydrocortisone cream). For inflammatory skin diseases.

Cantil (mepenzolate bromide). Antispasmodic for digestive disorders.

Capastat (capreomycin sulphate). Antituberculous antibiotic.

Capitol (benzalkonium). For dandruff.

Caplenal (allopurinol). For prevention of gout attacks.

Capoten (captopril). For hypertension.

Capozide (captopril with hydrochlorthiazide). For hypertension.

Capreomycin. Antituberculous antibiotic.

Caprin (aspirin for release in the intestine). Analgesic.

Captopril. For hypertension.

Carace (lisinopril). For high blood pressure.

Carbachol. Helps to relieve glaucoma or urinary retention.

Carbalax. Laxative suppositories.

Carbamazepine. For prevention of epilepsy and trigeminal neuralgia.

Carbaryl. Application for lice.

Carbellon. Antispasmodic based on belladonna.

Carbenicillin. Special-purpose penicillin.

Carbenoxolone. For peptic ulcer.

Carbidopa. Increases the activity of levodopa in Parkinson's disease.

Carbimazole. Controls hyperthyroidism.

Carbocisteine. For making sputum less viscous.

Carbo-Dome. Coal tar-containing application for psoriasis and eczema.

Carboplatin. Cisplatin-like anticancer drug.

Cardene (nicardipine hydrochloride). For prophylaxis and treatment of angina.

Cardiacap (pentaerythritol tetranitrate). For prophylaxis of angina.

Cardinol (propranolol). For high blood pressure and angina.

Cardura (doxazacin). Lowers blood pressure and probably cholesterol.

Carfecillin. Special-purpose penicillin for specific urinary infections.

Carisoma (carisoprodol). For skeletal muscle spasm.

Carisoprodol. For skeletal muscle spasm.

Carmellose. Protective coating for mouth ulcers.

Carmustine. Anticancer (alkylating) drug.

Carteolol. Beta-blocker eye drops for glaucoma.

Cascara. Stimulant laxative.

Castor oil. Stimulant laxative. Also a lubricant for eye drops.

Catapres (clonidine hydrochloride). For hypertension or control of migraine attacks.

Caved-S (deglycyrrhizinized liquorice). For peptic ulcer.

CCNU (lomustine). Anticancer drug.

Cedilanid (lanatoside C). Digitalis-like drug for heart failure and disorders of heart rhythm.

Cedocard (isosorbide dinitrate). For prophylaxis and treatment of angina.

Cefaclor. Cephalosporin antibiotic; especially for urinary infections.

Cefadroxil. Cephalosporin antibiotic similar to cefaclor.

Cefizox (ceftizoxime). Antibiotic similar to cefotaxime.

Cefotaxime. Newer cephalosporin; especially for urinary infections.

Cefoxitin. Cephamycin antibiotic, active against bowel bacteria.

Cefsulodin. Cephalosporin antibiotic, for surgical and burn infections.

Ceftazidime. Newer cephalosporin antibiotic similar to cefotaxime.

Ceftizoxime. Newer cephalosporin antibiotic similar to cefotaxime.

Cefuroxime. Newer, more penicillase-resistant, cephalosporin antibiotic.

Celbenin (methicillin). Early penicillinase-resistant penicillin.

Celevac (methylcellulose). Bulk laxative and appetite suppressant.

Cellucon (methylcellulose). Bulk laxative and appetite suppressant.

Centyl (bendrofluazide). Diuretic for hypertension or oedema.

Cephalexin. Cephalosporin antibiotic.

Cephalosporins. Antibiotics somewhat similar to the penicillins.

Cephalothin. Cephalosporin antibiotic less active than newer analogues.

Cephamandole. Cephalosporin antibiotic similar to cefuroxime.

Cephamycins. Antibiotics such as cefoxitin, similar to cephalosporins.

Cephazolin. Cephalosporin antibiotic.

Cephradine. Cephalosporin antibiotic.

Ceporex (cephalexin). Cephalosporin antibiotic.

Ceratonia. Antidiarrhoeal, adsorbent drug.

Cervagem (gemeprost). For dilatation of the uterine cervix.

Cesamet (nabilone). For nausea and vomiting in anticancer treatment.

Cetavlex (cetrimide). Antiseptic cream.

Cetavlon (cetrimide). Skin disinfectant.

Cetirizine. Antihistamine for allergies such as hay fever.

Cetriclens. Skin disinfectant solution containing chlorhexidine.

Cetrimide. Detergent antiseptic.

Cetylpyridinium. Antiseptic mouthrinse.

Chemotrim (co-trimoxazole). Antibacterial suspension for children.

Chendol (chenodeoxycholic acid). For gallstones.

Chenodeoxycholic acid. Given to dissolve cholesterol gallstones.

Chenofalk (chenodeoxycholic acid). For cholesterol gallstones.

Chloractil (chlorpromazine hydrochloride). Antipsychotic.

Chloral hydrate. For short treatment of insomnia, especially in children.

Chlorambucil. Anticancer drug.

Chloramphenicol. Toxic, reserve antibiotic for life-threatening infections unresponsive to other antibiotics. Also applied locally for eye or ear infections.

Chloraseptic. Phenol-containing antiseptic throat spray or gargle.

Chlorasol (sodium hypochlorite). Disinfectant solution.

Chlordiazepoxide. Early diazepam-like drug for anxiety.

Chlorhexidine. Effective disinfectant for skin and other purposes.

Chlormethiazole. Hypnotic drug for the elderly. Also given for alcohol withdrawal and, intravenously, for status epilepticus.

Chlormezanone. Anxiolytic and mild muscle relaxant drug.

Chloromycetin (chloramphenicol). Reserve antibiotic for specific life-threatening infections.

Chloroquine. For prophylaxis and treatment of malaria and an amoebicide.

Chlorothiazide. Diuretic for oedema or hypertension.

Chloroxylenol. Antiseptic.

Chlorpheniramine. Antihistamine for minor allergies.

Chlorpromazine. Tranquillizing antipsychotic drug and reserve anti-emetic.

Chlorpropamide. Oral antidiabetic drug.

Chlortetracycline. Tetracycline antibacterial.

Chlorthalidone. Diuretic for oedema or hypertension.

Chocovite. Vitamin D and calcium.

Cholecalciferol. Vitamin D.

Choledyl (choline theophyllinate). Bronchodilator for asthma and bronchitis.

Cholestyramine. Lowers blood cholesterol by impairing absorption.

Choline. Component of vitamin B complex.

Choline magnesium trisalicylate. Aspirin-like antirheumatic analgesic.

Choline salicylate. Application for mild inflammation of mouth or ears.

Choline theophyllinate. Bronchodilator tablets or syrup for asthma and bronchitis.

Cidomycin (gentamicin). Streptomycin-like antibiotic.

Cimetidine. Antihistamine which suppresses gastric acid secretion for treatment of peptic ulcer and persistent heartburn.

Cinnarizine. Antihistamine for motion sickness and Ménière's disease.

Cinobac (cinoxacin). Antibacterial.

Cinoxacin. Antibacterial drug for urinary tract infections.

Ciprofloxacin. Broad-spectrum antibacterial drug.

Ciproxin (ciprofloxacin). Antibacterial drug.

Cisapride. For acid reflux.

Cisplatin. Anticancer drug.

Claforan (cefotaxime). Cephalosporin antibiotic.

Clairvan (ethamivan). Post-operative respiratory stimulant.

Claradin. Effervescent aspirin.

Claritin (loratadine). Antihistamine for allergies such as hay fever.

Clavulanic acid. Added to penicillins to increase resistance to bacterial penicillinase.

Clemastine. Antihistamine for minor allergies.

Clindamycin. Antibiotic.

Clinicide. Preparation for elimination of body lice.

Clinoril (sulindac). Anti-inflammatory analgesic for rheumatic disorders.

Clioquinol. Weak antibacterial included in ear drops and skin preparations. Also included in antidiarrhoea mixtures but is neurotoxic.

Clobazam. Diazepam-like anxiolytic. Also has anti-epileptic activity.

Clobetasol propionate. Potent reserve corticosteroid for severe inflammatory skin diseases.

Clobetasone butyrate. Corticosteroid for inflammatory skin or eye disorders.

Clofazimine. Antibacterial drug for leprosy.

Clofibrate. Lowers blood cholesterol by interfering with its formation in the liver.

Clomid (clomiphene citrate). Anti-oestrogen for female infertility.

Clomiphene. Anti-oestrogen for some types of female infertility.

Clomipramine. Amitriptyline-like antidepressive drug.

Clomocycline. A tetracycline antibiotic.

Clonazepam. Diazepam-like anxiolytic. Injected for status epilepticus.

Clonidine. Antihypertensive drug.

Clopamide. Diuretic for oedema or hypertension.

Clopixol (zuclopenthixol). Long-acting depot injections for psychotic states including schizophrenia.

Clorazepate. Diazepam-like drug for anxiety states.

Clotrimazole. Antifungal drug for superficial infections.

Cloxacillin. Penicillinase-resistant penicillin for staphylococcal infections.

Coal tar preparations. For some types of psoriasis and eczema.

Cobadex. Hydrocortisone-containing cream for inflammations of the skin.

Cobalin-H (hydroxocobalamin, vitamin B_{12}) for pernicious anaemia.

Co-Betaloc (metoprolol tartrate with hydrochlorthiazide). For hypertension.

Cocaine (CD). Anaesthetic eye drops. Present in some analgesic mixtures for terminal care but now rarely used for this purpose.

Co-codamol (codeine with paracetamol). Analgesic.

Co-codaprin (codeine with aspirin). Analgesic.

Co-danthramer. Danthron-containing stimulant laxative. (now withdrawn).

Co-danthrusate. Danthron-containing stimulant laxative for the elderly only.

Codeine. In many preparations for analgesic, cough suppressant, or antidiarrhoeal actions.

Co-dergocrine. Possible adjunctive treatment for senile dementia.

Codis (codeine with aspirin). Analgesic.

Co-dydramol (di-hydrocodeine with paracetamol). Analgesic.

Co-fluampicil (flucloxacillin and ampicillin). Broad spectrum antibiotic mixture.

Cogentin (benztropine mesylate). For Parkinson's disease.

Colchicine. Alternative treatment for acute gout or short-term prophylaxis.

Colestid (colestipol). For hypercholesterolaemia.

Colestipol. Lowers blood cholesterol by impairing its absorption.

Colifoam (hydrocortisone preparation). For rectal inflammation.

Colistin. Polymyxin antibacterial drug mainly for skin, eye, and ear infections.

Colofac (mebeverine hydrochloride). For gastro-intestinal muscle spasm.

Cologel (methylcellulose). Bulk laxative.

Colomycin (colistin). Polymyxin antibacterial.

Colpermin. Peppermint oil capsules; for abdominal colic and distension.

Colven (mebeverine). Antispasmodic for digestive complaints.

Combantrin (pyrantel). For worm infestations.

Comox (co-trimoxazole). Broad-spectrum antibacterial drug.

Comploment continus. Vitamin B_6 for deficiency states.

Comprecin (enoxacin). For urinary tract and skin infections.

Concavit. Multivitamin preparations.

Concordin (protriptyline hydrochloride). Tricyclic anti-depressant.

Condyline (podophyllotoxin). For genital warts.

Congesteze (ephedrine and azatadine). Nasal decongestant.

Conova 30. Combined (ethinyloestradiol and ethynodiol diacetate) oral contraceptive.

Controvlar. Combined (oestrogen and progestogen) preparation for menstrual disorders.

Copholco. Pholcodine-containing cough linctus.

Co-proxamol (dextropropoxyphene hydrochloride and paracetamol). Analgesic.

Cordarone X (amiodarone hydrochloride). For disorders of heart rhythm.

Cordilox (verapamil). For disorders of heart rhythm, hypertension, or angina.

Corgard (nadolol). Beta-blocker for hypertension or angina.

Corgaretic (nadolol with bendrofluazide). For hypertension.

Corlan (hydrocortisone). Pellets for mouth ulcers.

Coro-Nitro Spray (glycerol trinitrate aerosol). For prophylaxis and treatment of angina.

Corsodyl (chlorhexidine gluconate). Antiseptic oral gel and mouthrinse.

Cortelan (cortisone acetate). For replacement therapy or for inflammatory or allergic diseases.

Cortenema (hydrocortisone enema). For inflammatory bowel disease such as ulcerative colitis.

Corticosteroids. Cortisone and other cortical steroids for replacement therapy, inflammatory and allergic diseases, allergic emergencies, and other purposes.

Corticotrophin. Hormone for stimulating release of natural corticosteroids.

Cortisol (hydrocortisone). A corticosteroid.

Cortisone acetate. Corticosteroid tablets.

Cortistab (cortisone acetate). Corticosteroid tablets.

Cortisyl (cortisone acetate). Corticosteroid tablets.

Corwin (xamaterol). For heart failure.

Cosalgesic (dextropoxyphene and paracetamol). For moderate pain or fever.

Cosmegen Lyovac (antinomycin D). Anticancer drug.

Cosuric (allopurinol). For prevention of attacks of gout.

Cotazym. Pancreatic digestive enzymes; for cystic fibrosis or post-pancreatectomy.

Co-trimoxazole (sulphamethoxazole and trimethoprim). Broad-spectrum antibacterial combination.

Cremalgin. Rubifacient counterirritant.

Creon. Pancreatic digestive enzymes; for pancreatic deficiency.

Crotamiton. Antipruritic skin applications.

Crystal violet paint. Traditional antiseptic application (use on broken skin not advised).

Crystapen (benzylpenicillin). Antibiotic injection.

Cuplex. Salicylate-containing application for inflammatory skin disorders.

Cyanocobalamin. Vitamin B_{12} for pernicious and related anaemias.

Cyclandelate. For peripheral blood vessel disease.

Cyclimorph (CD) (morphine with cyclizine). Analgesic and antinauseant.

Cyclizine. Antihistamine for nausea, vomiting, and vertigo.

Cyclobarbitone (CD). Barbiturate for intractable insomnia.

Cyclobral (cyclandelate). For peripheral blood vessel disease.

Cyclofenil. Anti-oestrogen for anovulatory infertility.

Cyclogest (progesterone). Suppositories for premenstrual syndrome.

Cyclopenthiazide. Diuretic for oedema or hypertension.

Cyclopentolate. Eye drops for dilating the pupils.

Cyclophosphamide. Cytoxic anticancer drug, sometimes also used for auto-immune diseases.

Cycloplegics. Drugs which paralyse the ciliary muscles of the eye and dilate the pupils, but raise intra-ocular pressure.

Cyclo-Progynova. Hormone combination for menopausal symptoms.

Cycloserine. Reserve antituberculous antibiotic.

Cyclospasmol (cyclandelate). For dilating peripheral blood vessels.

Cyclosporin. Selectively acting immunosuppressive drug.

Cyklokapron (tranexamic acid). For abnormal bleeding disorders.

Cyproheptadine. Antihistamine for minor allergies; also used to prevent migraine attacks.

Cyprostat (cyproterone acetate). Anti-androgen.

Cyproterone acetate. Anti-androgen. Second-line treatment for meta-stasizing prostatic cancer. Also used for aggressive behaviour in males or for acne and hirsutism in females.

Cytacon (cyanocobalamin, vitamin B_{12}). For pernicious and related anaemias.

Cytamen (cyanocobalamin). For pernicious and related anaemias.

Cytarabine. Anticancer drug for acute leukaemias.

Cytosar (cytarabine). Cytotoxic drug for acute leukaemias.

Cytotoxic agents. Drugs for anticancer treatment or immune suppression.

Dacarbazine. Reserve anticancer drug.

Dactinomycin (actinomycin D). Anticancer drug.

Dakin's solution (chlorinated soda solution). For wound irrigation.

Daktacort (miconazole with hydrocortisone). For inflammatory skin disorders with fungal infection.

Daktarin (miconazole). Antifungal drug.

Dalacin C (clindamycin). Reserve antibiotic.

Dalacin T (clindamycin). Local application for acne.

Dalivit. Multivitamin preparation.

Dalmane (flurazepam). Diazepam-like drug for insomnia.

Danazol. Hormone-release inhibitor for endometriosis and menstrual disorders.

Daneral SA (pheniramine maleate). Antihistamine mainly for minor allergies.

Danol (danazol). For endometriosis or for hereditary angio-oedema.

Danthron. Reserve stimulant purgative for the elderly only.

Dantrium (dantrolene sodium). For skeletal muscle spasm.

Dantrolene. Relaxes skeletal muscle spasm.

Daonil (glibenclamide). Oral drug for lowering blood sugar in diabetes.

Dapsone. Anti-leprosy drug. Also effective for some unrelated skin diseases.

Daranide (dichlorphenamide). For glaucoma.

Daraprim (pyrimethamine). For malaria prophylaxis or treatment.

Davenol. Pholcodine-containing cough suppressant linctus.

DDAVP (desmopressin). For antidiuretic hormone deficiency and for some haemorrhagic states.

Debrisoquine. Antihypertensive drug.

Decadron (dexamethasone). Corticosteroid for replacement therapy, or for inflammatory or allergic disorders.

Deca-Durabolin (nandrolone). For stimulating protein formation.

Decaserpyl (methoserpidine). Reserpine-like antihypertensive.

Declinax (debrisoquine). For hypertension.

Decortisyl (prednisone tablets). For replacement therapy or inflammatory and allergic diseases.

Dehydrocholic acid. Stimulates secretion of watery bile for washing out small stones.

Delfen (nonoxinol). Spermicidal contraceptive foam for use with barrier method.

Deltacortril Enteric, Delta-Phoricol, Deltastab. Prednisolone preparations for replacement therapy or for inflammatory and allergic diseases.

Demecarium bromide. Eye drops for glaucoma.

Demeclocycline hydrochloride. A tetracycline antibacterial.

Demulcents. Soothing substances purported to relieve dry, irritant cough.

De-Nol (bismuth chelate). For peptic ulcer.

Depixol (flupenthixol). Long-acting depot injection for schizophrenia.

Depo-Medrone (methylprednisolone acetate injection). For inflammatory and allergic disorders.

Deponit (glyceryl trinitrate skin preparation). For angina.

Depo-Provera (medroxyprogesterone acetate). Progestogen for menstrual disorders and reserve treatment for breast cancer.

Depostat (gestronol hexanoate). Progestogen; reserve treatment for breast or uterine cancers.

Dequadin (dequalinium chloride). Antiseptic mouthrinse and lozenges.

Dequalinium. Antiseptic mouthrinse and lozenges.

Derbac-C (carbaryl). Shampoo for head lice.

Derbac-M (malathion). Application for lice or scabies.

Dermovate (clobetasol propionate). Potent corticosteroid anti-inflammatory skin preparations.

Dermovate NN (clobetasol with neomycin). For infected inflammatory skin diseases.

Deseril (methysergide). For prophylaxis or migraine (specialist use only).

Desipramine. Tricyclic antidepressant.

Desmopressin. For antidiuretic hormone deficiency and for some haemorrhagic states.

Desogestrel. Progestogen; component of oral contraceptives.

Desonide. Potent corticosteroid for inflammatory and allergic skin diseases.

Desoxymethasone. Potent corticosteroid for inflammatory and allergic skin diseases.

Destolit (ursodeoxycholic acid). For dissolution of cholesterol gallstones.

Detoclo. Combination of three tetracycline antibacterial drugs.

Dettol (chloroxylenol). Antiseptic.

Dexamethasone. Corticosteroid; for suppressing inflammatory and allergic diseases.

Dexamphetamine sulphate (CD). For treatment of narcolepsy (earlier used for suppression of appetite).

Dexa-Rhinaspray (dexamethasone and neomycin). Spray for nasal allergy and inflammation.

Dexedrine (CD) (dexamphetamine sulphate). For narcolepsy.

Dextromethorphan. Opioid cough suppressant.

Dextromoramide (CD). Morphine-like analgesic.

Dextropropoxyphene. Minor analgesic derived from methadone.

DF 118 (dihydrocodeine tartrate). Opioid analgesic for moderate pain.

DHC Continus (dihydrocodeine tartrate). Opioid analgesic.

Diabinese (chlorpropamide). Oral drug for diabetes.

Diamicron (gliclazide). Oral drug for lowering blood sugar in diabetes.

Diamorphine (CD). Heroin; potent opioid analgesic and cough suppressant.

Diamox (acetazolamide). Diuretic to lower intra-ocular pressure in glaucoma.

Dianette (cyproterone). Anti-androgen for acne.

Diarrest (codeine phosphate and dicyclomine hydrochloride). Anti-diarrhoeal mixture.

Diatensec (spironolactone). Diuretic for specialist use.

Diazepam. Benzodiazepine anxiolytic, hypnotic, and lessens skeletal muscle spasm. Injection for control of status epilepticus.

Diazoxide. Emergency treatment for hypertensive crisis.

Dibenyline (phenoxybenzamine hydrochloride). For lowering blood pressure due to phaeochromocytoma and some other medical emergencies.

Dichloralphenazone. Chloral hydrate derivative for insomnia.

Dichlorphenamide. Lowers intra-ocular pressure in glaucoma.

Diclofenac. Non-steroidal anti-inflammatory analgesic for rheumatic diseases.

Diconal (CD) (dipipanone hydrochloride). Opioid analgesic.

Dicyclomine. Gastro-intestinal antispasmodic.

Dicynene (ethamsylate). For control of menstrual and some other bleeding disorders.

Didronel (disodium etidronate). For Paget's disease of bone.

Dienoestrol cream. Oestrogen application for postmenopausal vaginal disorders.

Diethylcarbamazine. Anthelmintic for tropical worm infestations.

Diethylpropion (CD). Appetite suppressant but addictive.

Difflam. Anti-inflammatory mouthrinse or cream.

Diflucan (fluconazole). For thrush.

Diflunisal. Aspirin derivative analgesic.

Digibind. For overdosage of digoxin.

Digitoxin. Digitalis analogue for heart failure and arrhythmias.

Digoxin. Similar to digitoxin but shorter action.

Dihydergot (dihydroergotamine mesylate). For migraine attacks.

Dihydrocodeine. Moderately potent opioid analgesic.

Dihydroergotamine. For control of migraine attacks.

Dihydrotachysterol. A form of vitamin D.

Dijex (aluminium hydroxide–magnesium carbonate). For dyspepsia.

Diloxanide. Amoebicide.

Diltiazem. For prophylaxis and treatment of angina.

Dimelor (acetohexamide). Oral drug for lowering blood sugar in diabetes.

Dimenhydrinate. Antihistamine for nausea and motion sickness.

Dimethicone. Sometimes included in antacids, purportedly to reduce flatulence.

Dimethindene. Antihistamine for minor allergies.

Dimotane (brompheniramine maleate). Antihistamine for minor allergies.

Dimotapp. Brompheniramine-containing nasal decongestant.

Dimyril (isoaminile citrate). For control of coughing.

Dindevan (phenindione). Anticoagulant for preventing thromboses.

Dinoprost and dinoprostone. Prostaglandins for induction of abortion.

Dioctyl (docusate sodium). Stimulant laxative but also included in ear wax preparations.

Dioderm (hydrocortisone cream). For inflammatory and allergic skin disorders.

Diovol. Antacid for dyspepsia.

Dipentum (olsalazine). For ulcerative colitis.

Diphenoxylate. Adjunctive treatment for diarrhoea.

Diphenylbutylpiperidines. A group of antipsychotic drugs used for schizophrenia.

Diphenylpyraline. Antihistamine for minor allergies.

Dipipanone (CD). Opioid analgesic.

Dipivefrine. Eye drops for dilating the pupil and for glaucoma.

Dipyridamole. For reducing the risk of thromboses.

Dirythmin SA (disopyramide). For control of disordered heart rhythm.

Disadine DP (povidone iodine). Skin disinfectant spray.

Disalcid (salsalate). Aspirin-like antirheumatic drug.

Disipal (orphenadrine hydrochloride). For Parkinson's disease.

Disodium cromoglycate (sodium cromoglycate). For prevention of asthmatic attacks.

Disopyramide. For control of irregularities of heartbeat.

Dispray (chlorhexidine). Skin disinfectant.

Disprol (paracetamol suspension). Analgesic.

Distaclor (cefaclor). Cephalosporin antibiotic.

Distalgesic (dextropropoxyphene hydrochloride and paracetamol). Analgesic tablets.

Distamine (penicillamine). For intractable rheumatoid arthritis.

Distaquaine (phenoxymethylpenicillin). Antibiotic tablets or syrup.

Distigmine. For muscle weakness in myasthenia gravis or some types
-- of urinary retention.

Disulfiram. Adjunctive treatment for alcoholism.

Dithranol. For psoriasis.

Dithrocream. Dithranol-containing cream; for psoriasis.

Dithrolan (dithronol with salicylic acid). For psoriasis.

Diumide-K Continus (frusemide). Diuretic with potassium for oedema or
hypertension.

Diuresal (frusemide). Diuretic for oedema or hypertension.

Diuretics. Increase urinary excretion of water and sodium for lessening
hypertension and the oedema of heart failure or other causes.

Diurexan (xipamide). Diuretic for oedema and hypertension.

Dixarit (clonidine). For prevention of migraine attacks.

Dobutamine. May stimulate the heart in states of circulatory shock.

Dobutrex (dobutamine). For circulatory shock.

Docusate sodium. Stimulant laxative. Also included in ear wax
preparations.

Dolmatil (sulpiride). Antipsychotic drug used for schizophrenia.

Dolobid (diflunisal). Aspirin-derivative, antirheumatic drug.

Doloxene (dextropropoxyphene hydrochloride). Opioid analgesic for mild
pain.

Domical (amitriptyline hydrochloride). Tricyclic antidepressant.

Domperidone. Anti-emetic; for nausea due to cytotoxic drugs.

Dopamet (methyldopa). For hypertension.

Dopamine. Adjunctive treatment of cardiogenic shock.

Dopram (doxapram hydrochloride). For post-operative respiratory
stimulation.

Doralese (indoramin). For prostatic obstruction.

Dormonoct (loprazolam). Diazepam-like drug for insomnia.

Dothiapin. Tricyclic antidepressive.

Double Check (nonoxinol). Spermicide for use with barrier method.

Doxapram. For post-operative respiratory stimulation.

Doxasacin. Lowers blood pressure and probably cholesterol.

Doxepin. Antidepressant.

Doxorubicin. Anticancer drug.

Doxycycline. A tetracycline antibiotic.

Dozic (haloperidol). Antipsychotic drug used for schizophrenia.

Dramamine (dimenhydrinate). Antihistamine for nausea and motion sickness.

Droleptan (droperidol). Antipsychotic drug used for schizophrenia and mania.

Dromoran (CD) (levorphanol tartrate). Morphine-like analgesic.

Droperidol. Antipsychotic sedating drug and anti-emetic used for schizophrenia and mania.

Drostanolone. Androgen; reserve treatment for breast cancer metastases.

Dryptal (frusemide). Diuretic for oedema or hypertension.

DTIC-Dome (dacarbazone). Reserve anticancer drug.

Dubam. Rubifacient counterirritant.

Dulcodos. Bisacodyl-containing stimulant laxative.

Dulcolax. Bisacodyl-containing laxative suppositories.

Duo-Autohaler (isoprenaline and phenylephrine). Bronchodilator aerosol for asthma and bronchitis.

Duofilm. Salicylic acid-containing preparation for warts.

Duogastrone (carbenoxalone sodium). For duodenal ulcer.

Duovent (fenoterol and ipratropium). Bronchodilator for asthma and bronchitis.

Duphalac (lactulose). Osmotic laxative.

Duphaston (dydrogesterone). For menstrual disorders.

Durabolin (nandrolone). For increasing post-operative protein formation.

Duracream (nonoxinol). Spermicidal contraceptive for use with barrier.

Duragel (nonoxinol). Spermicidal contraceptive for use with barrier.

Duromine (phentermine). Appetite suppressant.

Duvadilan (isoxsuprine hydrochloride). For ameliorating cerebral and peripheral vascular disease.

Dyazide (triamterine and hydrochlorothiazide). Diuretic combination for oedema and hypertension.

Dydrogesterone (progestogen). For menstrual disorders.

Dynese (magaldrate). For dyspepsia.

Dyspamet (chewable cimetidine). For peptic ulcer.

Dytac (triamterene). Potassium-sparing diuretic for oedema.

Dytide (triamterene and benzothiazide). Combined diuretic for oedema and hypertension.

Ebufac (ibuprofen). Anti-inflammatory analgesic.

Econacort (hydrocortisone with econazole). For inflammatory and fungal skin disorders.

Econazole. Antifungal antibiotic for skin or vaginal application.

Econocil VK (penicillin V tablets and capsules). Antibiotic.

Econmycin (tetracycline). Antibiotic.

Ecostatin (econazole nitrate). Antifungal skin or genital application.

Ecothiopate iodide. For relief of glaucoma.

Eczederm. Calamine-containing emollient cream for eczema.

Edecrin (ethacrynic acid). Potent diuretic for oedema.

Efcortelan (hydrocortisone ointment). For inflammatory and allergic skin disorders.

Efcortelan Soluble (hydrocortisone). Corticosteroid injection.

Efcortesol (hydrocortisone). Injection for allergic and other emergencies.

Effercitrate (potassium citrate). Tablets for mild urinary infections.

Effico. Vitamin mixture.

Efudix (fluorouracil). Antimetabolite anticancer drug.

Elantan (isosorbide mononitrate). For prophylaxis or treatment of angina.

Eldepryl (selegiline hydrochloride). Adjunctive treatment of Parkinson's disease.

Eldisine (vindesine sulphate). Vinca alkaloid anticancer drug.

Eltroxin (thyroxine sodium). For hypothyroidism.

Eludril. Chlorhexidine-containing antiseptic mouthwash and spray.

Elyzol (metronidazole). Antibacterial suppositories.

Emblon (tamoxifen). For breast cancer.

Emcor (bisoprolol). For high blood pressure and angina.

Emeside (ethosuximide). For absence seizures.

Emetine. Earlier used for amoebic infections.

Emetrol. Antispasmodic mixture; for dyspepsia.

Eminase (anistreplase). For dissolving coronary thromboses.

Enalapril. Antihypertensive drug.

En-De-Kay. Fluoride preparations for prevention of dental decay.

Endoxana (cyclophosphamide). Anticancer drug.

Enduron (methyclothiazide). Diuretic for oedema or hypertension.

Engerix B. Recombinant hepatitis B vaccine.

Enoxacin. For urinary tract and skin infections.

Enoximone. For heart failure.

Entamizole (diloxanide furoate). For chronic amoebic infections.

Enteromide (calcium sulphaloxate). Poorly absorbed sulphonamide antibacterial.

Epanutin (phenytoin). For epilepsy.

Ephedrine. Non-selective bronchodilator and nasal decongestant.

Ephynal (alpha-tocopherol). Vitamin E for specific deficiency diseases.

Epifrin. Adrenaline eye drops for open-angle glaucoma.

Epilim (sodium valproate). For all types of epilepsy.

Epirubicin. Anticancer drug.

Epodyl (ethoglucid). Alkylating anticancer drug.

Epogam (gamolenic acid). Evening primrose oil for eczema.

Epoprostenol (prostacyclin). Antiplatelet injection to prevent thrombosis.

Eppy (adrenaline). Eye drops for open-angle glaucoma.

Epsom salts (magnesium sulphate). Osmotic laxative.

Equagesic (CD) (ethoheptazine citrate, meprobamate and aspirin). Analgesic (not recommended).

Equanil (meprobamate). For anxiety (not usually recommended).

Eradacin (acrosoxacin). Antibacterial for penicillin-resistant gonorrhoea.

Erevax. Rubella vaccine.

Ergocalciferol. Vitamin D_2 for rickets and related deficiency diseases.

Ergometrine. For stimulation of uterine contraction.

Ergotamine. For acute migraine unresponsive to simple analgesics.

Erwinase (crisantaspase). Anticancer drug for leukaemia.

Erycen (erythromycin). Antibiotic.

Erymax (erythromycin). Antibiotic.

Erythrocin (erythromycin). Antibiotic.

Erythrolar (erythromycin). Antibiotic.

Erythromycin. Antibiotic; alternative to penicillin.

Erythroped. Erythromycin preparations, including paediatric mixtures.

Esbatal (bethanidine sulphate). For hypertension.

Esidrex (hydrochlorothiazide). Diuretic for oedema or hypertension.

Eskamel (sulphur and resorcinol cream). For acne.

Eskornade (diphenylpyraline and phenylpropanolamine capsules). For nasal decongestion.

Estracyt (estramustine phosphate). Anticancer agent.

Estradurin (polyestradiol phosphate). Oestrogen for prostatic cancer.

Estramustine. Anticancer alkylating agent.

Estrapak (oestradiol skin patch). For menopausal symptoms.

Estropipate (piperazine oestrone sulphate). Oestrogen for menstrual disorders or menopausal symptoms.

Ethacrynic acid. Potent diuretic for oedema.

Ethambutol. Antituberculous drug.

Ethamivan. Surgical respiratory stimulant.

Ethamsylate. For menorrhagia or other bleeding from small vessels.

Ethanolamine oleate. Injection for sclerosing varicose veins.

Ethinyloestradiol. Oestrogen for menstrual disorders and menopausal symptoms. Adjunctive treatment for some cases of breast cancer. Constituent of some combined oral contraceptives.

Ethionamide. Reserve antituberculous drug.

Ethoglucid. Anticancer drug.

Ethosuximide. For absence seizures.

Ethynodiol diacetate. Progestogen component of some oral contraceptives.

Etodolac. Anti-inflammatory analgesic for rheumatoid arthritis.

Etoposide. Anticancer drug.

Etretinate. For specialist use in severe psoriasis and other scaling skin diseases.

Etridronate disodium. For lowering blood calcium and for Paget's and some other bone diseases.

Eudemine (diazoxide). For hypertensive crises or chronic hypoglycaemia due to overproduction of insulin.

Euglucon (glibenclamide). Oral drug for lowering blood sugar in diabetes.

Eugynon (ethinyloestradiol and levonorgestrel). Combined oral contraceptive.

Eumovate (clobetasone butyrate). Corticosteroid-containing preparations for inflammatory and allergic skin diseases.

Eurax (crotamiton). Antipruritic application.

Eusol (chlorinated lime and boric acid solution). For wound irrigation.

Evacalm (diazepam). For short-term treatment of anxiety states.

Evadyne (butriptyline). Tricyclic depressant.

Evoxin (domperidone). Anti-emetic.

Exelderm (sulconazole nitrate). Antifungal application.

Exirel (pirbuterol). Bronchodilator for asthma and bronchitis.

Exolan (dithranol triacetate cream). For psoriasis.

Expulin (pseudoephedrine and chlorpheniramine). Compound cough preparation.

Expurhin (ephedrine and chlorpheniramine). Paediatric linctus nasal decongestant.

Fabahistin (mebhydrolin). Antihistamine for minor allergies.

Fabrol (acetylcysteine). For reducing sputum viscosity.

Famotidine. Cimetidine-like drug for peptic ulcer.

Fansidar (pyrimethamine with sulfadoxine). For prevention or treatment of malaria.

Farlutal (medroxyprogesterone acetate). Reserve treatment for breast cancer. Also used for endometrial and childhood renal cancer.

Fasigyn (tinidazole). For trichomonal vaginitis and anaerobic infections.

Faverin (fluvoxamine maleate). Antidepressant.

Fectrim (co-trimoxazole). Broad-spectrum antibacterial agent.

Fefol (ferrous sulphate and folic acid capsules). For anaemia. Also with zinc (*Fefol Z*) or mixed vitamins (*Fefol-Vit.*)

Feldene (piroxicam). Antirheumatic analgesic.

Fenbufen. Anti-inflammatory analgesic for rheumatic disease.

Fenfluramine. For suppressing appetite.

Fenoprofen. Anti-inflammatory analgesic for rheumatic disorders.

Fenopron (fenoprofen). Anti-inflammatory analgesic.

Fenostil (dimethindene maleate). Antihistamine for minor allergies.

Fenoterol. Bronchodilator inhalation for asthma and bronchitis.

Fentazin (perphenazine). Antipsychotic drug.

Feospan (ferrous sulphate slow release capsules). For pregnancy anaemia.

Ferfolic SV (folic acid and ferrous gluconate). For pregnancy anaemia.

Fergon (ferrous gluconate). For anaemia.

Ferrocap-F 350 (ferrous fumarate and folic acid). For pregnancy anaemia.

Ferrocontin Continus (ferrous glycine sulphate). For anaemia.

Ferrocontin Folic Continus (ferrous glycine sulphate with folic acid). For pregnancy anaemia.

Ferrograd (ferrous sulphate). For anaemia.

Ferrograd C (ferrous sulphate with vitamin C). For anaemia.

Ferrograd Folic (ferrous sulphate with folic acid). For pregnancy anaemia.

Ferromyn (ferrous succinate elixir). For anaemia.

Ferrous fumarate, gluconate, glycine sulphate and succinate. Iron salts, sometimes better tolerated than ferrous sulphate, for iron deficiency anaemia.

Fersaday (ferrous fumarate). For anaemia.

Fersamal (ferrous fumarate). For anaemia.

Fertiral (gonadorelin). Hormonal diagnostic agent.

Fesovit (ferrous sulphate with mixed vitamins). For deficiency diseases.

Finalgon. Rubifacient counterirritant.

Flagyl (metronidazole). Antibacterial and antiparasitic.

Flamazine (silver sulphadiazine). Antibacterial for skin infections.

Flavoxate. For urinary incontinence and frequency.

Flecainide. For disorders of heart rhythm.

Flolan (epoprostenol). Antithrombotic agent for renal dialysis.

Florinef (fludrocortisone acetate). Replacement therapy for Addison's disease.

Floxapen (flucloxacillin). Penicillinase-resistant penicillin.

Flu-Amp (flucloxacillin with ampicillin). Broad-spectrum antibiotic combination.

Fluanxol (flupenthixol). For depression and psychoses.

Fluclorolone. Potent corticosteroid for inflammatory skin diseases.

Flucloxacillin. Penicillinase-resistant penicillin for staphylococcal infections.

Flucytosine. Antifungal drug for systemic fungal infections.

Fludrocortisone. Mineralocorticosteroid for adrenal insufficiency (Addison's disease).

Flumazenil. Antagonist of benzodiazepines such as diazepam.

Flunisolide. Corticosteroid spray for nasal allergies.

Flunitrazepam. Diazepam-like hypnotic.

Fluocinolone. Potent corticosteroid application for inflammatory skin diseases.

Fluocinonide. Potent corticosteroid for severe inflammatory skin diseases.

Fluocortolone. Corticosteroid application for inflammatory skin diseases.

Fluor-a-day Lac. Fluoride tablets for prevention of dental decay.

Fluoride preparations. For prevention of dental decay.

Fluorigard. Fluoride preparations for prevention of dental decay.

Fluorometholone. Corticosteroid eye drops for inflammatory conditions.

Fluorouracil. Anticancer drug.

Flupenthixol. Antidepressive and antipsychotic drug.

Fluphenazine. Antipsychotic depot preparations.

Flurandrenolone. Corticosteroid application for skin inflammations.

Flurazepam. Diazepam-like hypnotic for short-term treatment of insomnia.

Flurbiprofen. Anti-inflammatory analgesic for rheumatic diseases.

Fluspirilene. Antipsychotic for schizophrenia.

Fluvirin. Influenza vaccine.

Fluvoxamine. Antidepressant.

FML (fluorometholone). Corticosteroid eye drops.

Folex-350 (folic acid and ferrous fumarate). For pregnancy anaemia.

Folic acid. For correction of folate-deficiency anaemia.

Folicin (ferrous sulphate and folic acid). For pregnancy anaemia.

Folinic acid. Antidote to methotrexate and other folate antagonist anticancer drugs.

Follicle-stimulating hormone (FSH). For infertility secondary to pituitary deficiency.

Forceval (ferrous fumarate with mixed vitamins and minerals). For deficiency diseases.

Formaldehyde. For treatment of warts.

Formulix (paracetamol and codeine elixir). For pain or coughs in children.

Fortral (CD) (pentazocine). Morphine-like analgesic tablets and injection.

Fortum (ceftazidime). A cephalosporin antibiotic.

Fortunan (haloperidol). For psychotic disorders.

Fosfestrol. Oestrogen for prostatic cancer.

Framycetin. Streptomycin-like antibiotic for local application.

Framycort (framycetin and hydrocortisone). For inflammatory skin, eye, and ear disorders.

Framygen (framycetin). Antibiotic for skin, eye, and ear infections.

Franol (CD) (ephedrine, theophylline, and phenobarbitone). Bronchodilator for asthma (not usually recommended).

Frisium (clobazam). Diazepam-like anxiolytic and adjunctive treatment of epilepsy.

Froben (flurbiprofen). Anti-inflammatory analgesic for rheumatic diseases.

Frumil (amiloride with frusemide). Potassium-sparing diuretic for oedema.

Frusemide. Potent diuretic for oedema.

Frusene (triamterene with frusemide). Potassium-sparing diuretic for oedema.

Frusetic (frusemide). Potent diuretic for oedema.

Frusid (frusemide). Potent diuretic for oedema.

FSH. Follicle-stimulating hormone for pituitary deficiency infertility.

Fucibet (betamethasone with fusidic acid). Anti-inflammatory skin cream.

Fucidin (sodium fusidate). Antistaphylococcal antibiotic.

Fucidin H (hydrocortisone with fusidic acid). For inflammatory and infected skin disorders.

Fucithalmic (fusidic acid). For staphylococcal eye infections.

Fulcin (griseofulvin). Oral drug for ringworm of the skin and nails.

Fungilin (amphotericin). Antifungal lozenges and suspension.

Fungizone (amphotericin injection). For systemic fungal infections.

Furadantin (nitrofurantoin). Antibacterial for urinary infections.

Fusidic acid and sodium fusidate. Two forms of an anti-staphylococcal antibiotic.

Fybogel (ispaghula husk). Bulk-forming laxative.

Fybranta (bran). Bulk-forming laxative.

Galcodine (codeine). Cough suppressant linctus.

Galenphol (pholcodine). Cough suppressant linctus.

Galfer (ferrous fumarate). For iron deficiency.

Galfer FA (ferrous fumarate with folic acid). For pregnancy anaemia.

Galfer-Vit (ferrous fumarate with mixed vitamins). For deficiency diseases.

Galpseud (pseudoephedrine). Nasal decongestant.

Gamanil (lofepramine). Tricyclic antidepressant.

Gamolenic acid. Evening primose oil for eczema.

Ganda (guanethidine monosulphate eye drops). For glaucoma.

Garamycin (gentamicin drops). For eye or ear infections.

Gardenal sodium (CD) (phenobarbitone). Antiepileptic.

Gastrese (metoclopramide). For nausea and vomiting.

Gastrils. Antacid for dyspepsia.

Gastrobid Continus (long-acting metoclopramide hydrochloride). For nausea and vomiting.

Gastrocote. Antacid for dyspepsia.

Gastromax (long-acting metoclopramide hydrocloride). Anti-emetic.

Gastron. Antacid for indigestion.

Gastrozepin (pirenzepine). For peptic ulceration.

Gaviscon. Antacid for indigestion.

Gee's linctus. Opium and squill cough linctus.

Gelcosal (coal tar gel). For eczema and psoriasis.

Gelcotar (coal tar gel). For eczema and psoriasis.

Gelusil. Antacid for indigestion.

Gemfibrozil. For lowering blood cholesterol levels.

Gentamicin. Streptomycin-like antibiotic.

Genticin (gentamicin). Antibiotic.

Genticin HC (hydrocortisone and gentamicin). Anti-inflammatory and antibacterial preparations for skin.

Gentisone HC (gentamicin and hydrocortisone). Anti-inflammatory and antibacterial ear drops.

Gestanin (allyloestrenol). Progestagen for habitual abortion.

Gestodene. Progestogen in some combined oral contraceptives.

Gestone (progesterone). Mainly for menstrual disorders.

Gestronol. A progestogen; reserve treatment for breast and uterine cancer.

Givitol (ferrous fumarate with mixed vitamins). For anaemia and other deficiency diseases.

Glandosane. Artificial saliva spray for dry mouth.

Glauline (metipranolol). Beta-blocker eye drops for glaucoma.

Gliclazide. Oral drug for lowering blood sugar in diabetes.

Glipizide. Oral drug for lowering blood sugar in diabetes.

Gliquidone. Oral drug for lowering blood sugar in diabetes.

Glucagon. Emergency injection for hypoglycaemic (insulin) coma.

Glucocorticoids. Corticosteroids mainly given for inflammatory and allergic diseases.

Glucophage (metformin hydrochloride). Oral drug for lowering blood sugar in diabetes.

Glucotard (guar gum). Retards sugar absorption from carbohydrates.

Glurenorm (gliquidone). Oral treatment of diabetes mellitus.

Glutaraldehyde. Skin application for warts.

Glutarol (glutaraldehyde). For warts.

Glyceryl trinitrate. For prophylaxis and treatment of angina.

Glyconon (tolbutamide). Oral drug for lowering blood sugar in diabetes.

Glycopyrronium. Antispasmodic for gastro-intestinal disorders.

Glykola (ferric chloride preparation). For anaemia.

Glymese (chlorpropamide). Oral drug for lowering blood sugar in diabetes.

Glymidine. Oral drug for lowering blood sugar in diabetes.

Glypressin (terlipressin). For reducing bleeding from oesophageal varices.

Gold salts. Injection or tablets for amelioration of rheumatoid arthritis.

Gonadotraphon LH (chorionic gonadotrophin). Adjunctive treatment of pituitary deficiency infertility.

Goserelin. Hormone antagonist for treatment of prostatic cancer.

Graneodin (neomycin sulphate). Antibacterial skin and eye ointments.

Gregoderm (hydrocortisone, neomycin, nystatin, and polymixin B). Anti-inflammatory skin ointment.

Griseofulvin. Taken by mouth for ringworm.

Grisovin (griseofulvin tablets). For ringworm of the skin and nails.

Growth hormone (somatrem). For treatment of short stature due to pituitary deficiency.

GTN (glyceryltrinitrate). For prophylaxis and treatment of angina.

Guanethidine. For adjunctive treatment of hypertension.

Guanor Expectorant. Compound syrup for relief of cough.

Guar gum. Fibre for retarding sugar absorption from carbohydrate or for providing bulk for intestinal disorders.

Guarem and *Guarina* (guar gum). To provide fibre for bowel and other disorders.

Gynatren (inactivated *Lactobacillus* preparation). For recurrent vaginal trichomoniasis.

Gyno-Daktarin (miconazole application). For genital candidosis (thrush).

Gynol II (nonoxinol). Spermicidal contraceptive for use with barrier.

Gyno-Pevaryl (econazole cream and pessaries). For genital candidosis (thrush).

Haelan (flurandrenolone). Corticosteroid preparations for inflammatory skin disorders.

Halciderm (halcinonide). Potent corticosteroid cream for inflammatory skin disorders.

Halcinonide. Potent anti-inflammatory corticosteroid for local application.

Halcion (triazolam). Diazepam-like hypnotic for short-term treatment of insomnia.

Haldol (haloperidol). Tranquillizing antipsychotic agent used for schizophrenia.

Half-Inderal LA (long-acting propranolol). For hypertension and angina.

Haloperidol. Tranquillizing antipsychotic drug used for schizophrenia.

Halycitrol. Vitamins A and D for deficiency states.

Hamamelis suppositories. For symptomatic relief of piles.

Hamarin (allopurinol). For prophylaxis of gout.

Harmogen (estropipate). Oestrogen for menstrual and menopausal disorders.

Haymine (ephedrine and chlorpheniramine). Nasal decongestant.

H-B-Vax. Hepatitis B vaccine.

HCG. Human chorionic gonadotrophin. Adjunctive treatment of hypopituitary infertility.

Heminevrin (chlormethiazole). Hypnotic. Also given for status epilepticus and alcohol withdrawal symptoms.

Heparin. Short-acting anticoagulant injection for prevention of thrombosis.

Hep-Flush, Heplok, Hepsal. Heparin for prevention of thromboses.

Heroin (CD) (diamorphine). Potent morphine-derived analgesic and anxiolytic.

Herpid (idoxuridine). Antiviral drug for herpetic infections of the skin.

Hexachlorophane. Skin disinfectant.

Hexamine (methenamine). Antibacterial drug for urinary infections.

Hexetidine. Antiseptic in mouthrinses.

Hexopal (nicotinic acid derivative). For poor peripheral circulation.

Hibidil and Hibisol (chlorhexidine). Antiseptic solutions.

Hibiscrub (chlorhexidine). Surgical skin disinfectant.

Hibitane (chlorhexidine). Antiseptic for skin and for controlling dental plaque.

Hioxyl (hydrogen peroxide). Antiseptic cream.

Hiprex (hexamine). Antibacterial for urinary infection.

Hismanal (astemizole). Antihistamine for minor allergies.

Histalix. Compound cough preparation.

Histryl (diphenylpyraline hydrochloride). Antihistamine for minor allergies.

Homatropine. Eye drops for dilating the pupils.

Honvan (fosfestrol tetrasodium). Oestrogen precursor for treatment of prostatic cancer.

Hormofemin (dienoestrol). Oestrogen cream for vaginal and vulval disorders.

Hormonin (oestradiol). For menstrual and menopausal disorders.

Human insulins. Insulins produced by recombinant DNA techniques, to reduce the risk of insulin allergy.

Humotet. Immunoglobulin for tetanus prophylaxis.

Humulin. Human insulins.

Hydergine (co-dergocrine mesylate). For poor peripheral circulation.

Hydralazine. Antihypertensive drug; usually an adjunctive treatment.

Hydrea (hydroxyurea). Anticancer drug mainly for chronic leukaemia.

Hydrenox (hydroflumethiazide). Diuretic for oedema or hypertension.

Hydrochlorothiazide. Diuretic for oedema or hypertension.

Hydrocortisone. Active form of cortisone for replacement of deficiency or for inflammatory and allergic disorders.

Hydrocortistab (hydrocortisone preparations). For replacement therapy or for inflammatory and allergic diseases.

Hydrocortisyl. Hydrocortisone ointment.

Hydrocortone (hydrocortisone tablets). For replacement treatment or inflammatory diseases.

Hydroflumethiazide. Diuretic for oedema or hypertension.

Hydrogen peroxide. Oxidizing antiseptic solution or mouthwash.

Hydromet (methyldopa with hydrochlorothiazide). For hypertension.

HydroSaluric (hydrochlorothiazide). Diuretic for oedema or hypertension.

Hydrotalcite (aluminium–magnesium compound antacid). For dyspepsia.

Hydroxyocobalamin (vitamin B_{12} injection). For pernicious anaemia.

Hydroxychloroquine. For prophylaxis or treatment of malaria, or for treatment of active rheumatoid arthritis and lupus erythematosus.

Hydroxycholecalciferol. A form of vitamin D.

Hydroxyprogesterone (a progestogen). Mainly for habitual abortion.

Hydroxyurea. Anticancer drug, mainly for chronic myeloid leukaemia.

Hydroxyzine. For anxiety states (not usually recommended).

Hygroton (chlorthalidone). Diuretic for oedema or hypertension.

Hyoscine hydrobromide (scopolamine). For control of nausea and vertigo, and for pre-operative medication.

Hypercal (rauwolfia alkaloids). For mild hypertension.

Hypertane 50 (amiloride with hydrochlorthiazide). Potassium-sparing diuretic for oedema.

Hypon. Aspirin-containing, compound analgesic.

Hypotears. Polyvinyl alcohol artificial tears for dry eyes.

Hypovase (prazosin hydrochloride). Antihypertensive drug.

Hypurin preparations. Isophane (long-acting) insulins given sub-cutaneously, usually with other insulins, for diabetes.

Hytrin (terazosin hydrochloride). For mild to moderate hypertension.

Ibular and *Ibumetin* (ibuprofen). Anti-inflammatory analgesic.

Ibuprofen. Anti-inflammatory analgesic for rheumatic diseases.

Ichthammol preparations. Applications for chronic eczema.

Idoxene (idoxuridine). Eye ointment for herpetic infections.

Idoxuridine. Antiviral drug for local application to herpetic infections.

Iduridin (idoxuridine in dimethyl sulphoxide). Skin application for herpes.

Ifosfamide. Cyclophosphamide-like anticancer drug.

Ilosone (erythromycin estolate). Antibiotic.

Ilube. Acetylcysteine-containing drops for dry eyes.

Imbrilon (indomethacin). Anti-inflammatory analgesic for rheumatic disorders.

Imdur (isosorbide mononitrate). Long-acting prophylaxis and treatment of angina.

Imferon (iron dextran injection). For iron-deficiency anaemia.

Imidazoles. Antifungal drugs, including miconazole, ketoconazole and others.

Imipramine. Tricyclic antidepressant.

Imodium (loperamide). For diarrhoea.

Imperacin (oxytetracycline). Antibiotic.

Imunovir (inosine pranobex). Adjunctive treatment of viral infections.

Imuran (azathioprine). Immunosuppressive drug.

Inderal (propranolol). For hypertension and angina.

Inderetic and *Inderex* (propranolol with bendrofluazide). For hypertension.

Indocid, Indoflex, Indolar, Indomod. Indomethacin preparations.

Indomethacin. Anti-inflammatory analgesic for rheumatic disorders.

Indoramin. Antihypertensive drug usually used in combination with others.

Infacol (dimethicone). For dyspepsia.

Initard preparations. Isophane (long-acting) insulins for diabetes.

Innovace (enalapril maleate). For hypertension.

Inosine pranobex. Adjunctive treatment of viral infections.

Inositol nicotinate. For poor peripheral circulation.

Insulatard. Isophane (long-acting) insulins.

Insulin. Given by injection to lower blood sugar in diabetes.

Intal (sodium cromoglycate). Used to prevent asthmatic attacks.

Integrin (oxpertine). Similar to chlorpromazine; for schizophrenia and other excited psychotic states.

Interferons. Used for rare form of leukaemia and for AIDS.

Intralgin. Local rubefacient.

Intron A (interferon for rare form of leukaemia and for AIDS).

Intropin (dopamine). Used for emergency stimulation of heart.

Ipral (trimethoprim). Antibacterial used in urinary infections.

Iprindole. Antidepressive, similar side-effects to amitriptyline, but less sedative.

Iron. Given for iron-deficiency anaemia.

Isocarboxazid (*Marplan*). Monoamine oxidase inhibitor for depression.

Isogel (ispaghula husk). Bulk laxative.

Isoket (isosorbide dinitrate). Used in angina pectoris.

Isoniazid. Antituberculosis drug.

Isoprenaline. Used for severe slowing of the heart and for asthma.

Isordic (isosorbide dinitrate). Used for angina pectoris.

Isosorbide mononitrate. Used for angina pectoris.

Isotrate (isosorbide mononitrate). For angina pectoris.

Isotretinoin (*Roaccutane*). Related to vitamin A, given orally for acne.

Ispaghula husk. Bulk purgative.

Isradipine. For high blood pressure.

Itraconazole. For thrush and other fungal infections.

Kalspare (triamterine with chlorthalidone). Diuretic.

Kalten (amiloride hydrochloride, hydrochlorthiazide, and atenolol). Antihypertensive.

Kanamycin. Streptomycin-like antibiotic.

Kaodene (kaolin and codeine phosphate). Symptomatic treatment of diarrhoea.

Kaolin. Included in antidiarrhoeal mixtures.

Kaopectate (kaolin). For diarrhoea.

Karvol. Inhalation for relief of nasal obstruction.

Kefadol (cephamandole). Antibiotic.

Keflex (cephalexin). Antibiotic.

Keflin (cephalothin). Antibiotic.

Kefzol (cephazolin). Antibiotic.

Kelfizine W (sulphometopyrazine). Antibacterial.

Kemadrin (procyclidine hydrochloride). For Parkinson's disease.

Kemicetine (chloramphenicol). Antibiotic.

Kenalog (triamcinolone). Long-acting steroid injection for hay fever.

Keratolytics. To remove scale from skin.

Kerecid (idoxuride). Eye drops for herpes infections.

Kerlone (betaxolol hydrochloride). Beta-blocker; many uses including hypertension.

Kest (phenolphthalein and magnesium sulphate). Laxative.

Ketazolam. Benzodiazepine for the short-term treatment of anxiety.

Ketoconazole. Antifungal.

Ketoprofen. Non-steroid anti-inflammatory analgesic for rheumatic disorders.

Ketotifen. Capsules for prevent asthmatic attacks.

Kiditard (quinidine). Regulates heart rhythm.

Kinidin Durules (quinidine). Regulates heart rhythm.

KLN (kaolin, pectin, and sodium citrate). Antidiarrhoeal.

Kloref. Potassium supplement.

Kolanticon (dicyclomine). Relieves spasm of intestine.

Labetalol. Beta-blocker used for hypertension.

Labophylline injection. Long-acting bronchodilator for asthma and bronchitis.

Laboprin. Aspirin.

Labosept. Antiseptic pastilles.

Labrocol (labetalol). Beta-blocker.

Lachesine. Eye drops to dilate pupil.

Lactulose. Liquid purgative.

Ladropen (flucloxacillin). Antibiotic.

Lanvis (thioguanine). For acute leukaemia.

Laractone (spironolactone). A diuretic reserved for specialist use.

Laraflex (naproxen). Anti-inflammatory analgesic.

Larapam (piroxicam). Anti-inflammatory analgesic.

Laratrim (co-trimoxazole). Antibacterial used in urinary and chest infections.

Largactil (chlorpromazine). Neuroleptic. Tranquillizer for some mental illnesses, including schizophrenia.

Lariam (mefloquine). For prophylaxis of malaria.

Larodopa (levodopa). Used in Parkinson's disease.

Lasikal (frusemide with potassium). Diuretic with potassium.

Lasilactone (frusemide with spironolactone). Diuretic mixture.

Lasipressin (penbutolol with frusemide). Beta-blocker and diuretic for hypertension.

Lasix (frusemide). Diuretic.

Lasix+K (frusemide with potassium). Diuretic with potassium.

Lasma (slow release theophylline). Bronchodilator for asthma and bronchitis.

Lasonil. Ointment for piles.

Lasoride (amiloride). Diuretic.

Lassar's paste (zinc and salicylic acid paste). Skin application for hard, scaly conditions.

Latamoxef. Antibiotic.

Laxoberal (sodium picosulphate). Purgative, used before bowel X-rays.

Ledercort (triamcinolone acetonide). Steroid tablets used for severe allergy and such other inflammatory diseases.

Lederfen (fenbufen). Anti-inflammatory analgesic used in rheumatism.

Ledermycin (demeclocycline hydrochloride). A tetracycline antibiotic.

Lederspan (triamcinolone hexacetonide). Steroid injection which can be put into a joint.

Lejfibre (bran biscuits). High-fibre food.

Lejguar (guar gum). Fibre taken with water during meals which slows absorption of sugar. Used in diabetes.

Lentard MC (insulin zinc suspension). Long-acting insulin injection for diabetes.

Lentizol (slow-release amitriptyline). Antidepressant.

Leo K (potassium chloride). Potassium supplement.

Lergoban (diphenylpyraline hydrochloride). Antihistamine for allergies.

Leukeran (chlorambucil). Anticancer, immunosuppressant.

Levamisole. Used for ascaris (round worm) infestations. Not available in UK.

Levius (slow-release aspirin). Aspirin which may have reduced potential to irritate the stomach.

Levodopa. Used in Parkinson's disease.

Levonorgestrel. Progesterone hormone, used in combined oral contraceptives and in progesterone-only contraceptive.

Levorphanol. Powerful opioid analgesic.

Lexotan (bromazepam). Benzodiazepine hypnotic and anxiolytic.

Lexpec (folic acid). Member of B group of vitamins. Given for some forms of anaemia.

LH-RH (gonadotrophin-releasing hormone). Used for diagnosis and for the treatment of infertility.

Libanil (glibenclamide). Oral antidiabetic drug.

Librium (chlordiazepoxide hydrochloride). Benzodiazepine for the short-term relief of anxiety and insomnia.

Lidoflazine. Calcium antagonist, used for angina.

Lignocaine. Local anaesthetic. Also used for abnormal heart rhythms.

Limbitrol (chlordiazepoxide with amitriptyline). Mixture of benzo-diazepine and antidepressant.

Lincocin (lincomycin). Antibiotic—very limited use.

Lindane. Local application for lice and scabies.

Lingraine (ergotamine tartrate). For migraine attacks.

Lioresal (baclofen). For muscle spasticity.

Liothyronine. One of the thyroid hormones, used in myxoedema.

Liquid paraffin. Lubricant purgative. Not recommended.

Liquorice, deglycyrrhizinized. Used for peptic ulceration.

Lisinopril. For high blood pressure.

Liskonum (lithium carbonate). Mood-stabilizing agent, used in some recurrent psychological illnesses.

Litatex (lithium citrate). Mood-stabilizing agent, used in some recurrent psychological illnesses.

Lithium. Mood-stabilizing agent, used in some recurrent psychological illnesses.

Lobak (chlormezonone with paracetamol). Compound analgesic tablet.

Locabiotal (fusafungine). Anti-infective nasal spray.

Locobase. Ointment for use in the skin.

Locoid (hydrocortisone butyrate in Locobase). Weak steroid application for skin.

Locoid C (hydrocortisone butyrate and chlorquinalol). Antibacterial and steroid application for skin.

Locorten-Vioform (flumethasone pivalate and clioquinol). Anti-infective and anti-irritant ear drops.

Lodine (etodolac). Non-steroidal anti-inflammatory analgesic for rheumatic disorders.

Loestrin (norethisterone acetate and ethinyloestradiol). Combined oral contraceptive.

Lofepramine. Antidepressive.

Logynon (levonorgestrel and ethinyloestradiol). Combined oral contraceptive.

Lomotil (atropine and diphenoxylate). Symptomatic treatment of diarrhoea.

Lomustine. Anticancer drug.

Loniten (minoxidil). Antihypertensive.

Loperamide. Symptomatic treatment of diarrhoea.

Lopid (gemfibrozil). Used in conjunction with diet to lower raised blood lipids.

Loprazolam. Benzodiazepine, mainly used to promote sleep.

Lopresor (metoprolol). Beta-blocker. Many uses, including hypertension.

Lopresoretic (metoprolol tartrate and chlorthalidine). Beta-blocker and diuretic mixture for hypertension.

Loratadine. Antihistamine for allergies such as hay fever.

Lorazepam. Benzodiazepine for the short-term treatment of anxiety and insomnia.

Lorexane (lindane). Cream and shampoo for scabies and lice.

Lormetazepam. Benzodiazepine, mainly used to promote sleep.

Losec (omeprazole). For peptic ulcers.

Lotussin (dextromethorphan, diphenhydramine, ephedrine, and guaiphenesin). Cough suppressant.

Ludiomil (maprotiline hydrochloride). Antidepressant.

Lugol's solution (iodine in potassium iodide). Pre-operative treatment for thyroid overactivity.

Lurselle (probucol). Lowers blood cholesterol levels.

Lymecycline. Tetracycline antibiotic.

Lynoestrenol. Female progestogen hormone in oral contraceptive.

Lypressin. Synthetic hormone analogue, used in vasopressin deficiency.
Lysuride. For Parkinson's disease.

Maalox (aluminium hydroxide and magnesium hydroxide). Antacid.
Macrodantin (nitrofurantoin). For urinary tract infections.
Madopar (levodopa and benserazide). Used in Parkinson's disease.
Mafenide. Antibacterial used for skin infections.
Magaldrate (combination of aluminium and magnesium hydroxides and sulphuric acid). Used in dyspepsia.
Magnapen (ampicillin and flucloxacillin). Powerful antibiotic mixture.
Magnesium carbonate. Antacid.
Magnesium hydroxide. Mild purgative and antacid.
Magnesium oxide. Mild purgative and antacid.
Magnesium sulphate. Saline purgative.
Magnesium trisilicate. Antacid.
Malarivon (chloroquine). Treatment and prevention of malaria.
Malathion. Local application for scabies and lice.
Malinal (almasilate). Used in dyspepsia.
Malix (glibenclamide). Oral antidiabetic.
Maloprim (pyrimethamine and dapsone). Treatment and prevention of malaria.
Maprotiline. Antidepressant.
Marevan (warfarin). Anticoagulant.
Marplan (isocarboxazid). Monoamine oxidase inhibitor used in depression.
Marvelon (ethinyloestradiol and desogestrel). Combined oral contraceptive.
Maxepa (omega-3 fish oil). Used to reduce abnormally raised blood lipids.
Maxidex (dexamethasone). Steroid eye drops for specialist use in some inflammatory conditions.
Maxitrol (dexamethasone with antibiotics). Eye drops for specialist use in some infected, inflamed conditions.
Maxolon (metoclopramide). Powerful antinauseant.
Maxtrex (methotrexate). Anticancer drug.
Mazindol. Amphetamine-like drug. Used for weight loss. Not recommended.
Mebendazole. For infestations with threadworm, hookworm, and roundworm.
Mebeverine. Antispasmodic, used in spastic colon.
Mebhydrolin. Antihistamine for allergies.

Mecillinam. Antibiotic.

Meclozine. Antinauseant, used in vertigo and travel sickness.

Medazepam (benzodiazepine). Used in short-term treatment of anxiety.

Medihaler-Duo (isoprenaline and phenylephrine aerosol). Inhaled for asthma and bronchitis.

Medihaler-epi (adrenalin aerosol). Inhaled for asthma and bronchitis.

Medihaler-Ergotamine (ergotamine aerosol). Inhaled for migraine.

Medihaler-iso (isoprenaline aerosol). Inhaled for asthma and bronchitis.

Medilave gel (benzocaine). Local anaesthetic gel for painful mouth lesions.

Medised (paracetamol and promethazine). Analgesic/sedative mixture.

Medocodene (paracetamol and codeine). Analgesic mixture.

Medomet (methyldopa). Antihypertensive.

Medrone acne lotion (sulphur, methylprednisolone acetate, and aluminium chlorhydroxide). Sulphur, steroid, and drying agent for acne.

Medrone tablets (methylprednisolone). Powerful steroid used in some inflammatory and allergic diseases.

Medroxyprogesterone. Progestogen hormone, used as contraceptive and in cancer treatment.

Mefenamic acid. Analgesic, used in rheumatic disease.

Mefloquine. For prophylaxis of malaria.

Mefoxin (cefoxitin). Antibiotic.

Mefruside. Thiazide diuretic, used in hypertension and oedema.

Megace (megestrol acetate). Progestogen hormone, used in cancer treatment.

Megaclor (clomocycline sodium). Tetracycline antibiotic.

Megestrol acetate. Progesterone-like hormone, used in cancer treatment.

Melleril (thioridazine). Neuroleptic tranquillizer, used in mental illness, particularly in the elderly.

Melphalan. Anticancer drug.

Menadiol. Water-soluble vitamin K.

Menotrophin. Hormone used to treat infertility.

Mepacrine. Used for giardia infections.

Mepenzolate bromide. Used for spasm of the intestinal tract.

Meptazinol. Powerful analgesic. Similar action to morphine.

Meptid (meptazinol). Powerful analgesic. Similar action to morphine.

Mequitazine. Antihistamine, for allergies.

Merbentyl (dicyclomine hydrochloride). Used for spasm of the intestinal tract.

Mercaptopurine. Anticancer drug.

Mercuric oxide. Has been used for eye infections. Not recommended.

Merocaine (benzocaine and cetylpyridinium chloride). Lozenges containing local anaesthetic and antiseptic.

Merocets (cetylpyridinium chloride). Lozenges and solution to clean mouth.

Mesalazine. Used in ulcerative colitis.

Mesterolone. Hormone with testosterone-like effects.

Mestinon (pyridostigmine bromide). Improves strength in myasthenia gravis.

Mestranol. Oestrogen used in combined oral contraceptive.

Metamucil (ispaghula husk). Bulk agent, used in constipation and to regulate colostomies.

Metanium (titanium ointment). Drying agent for napkin rash and moist lesions in skin folds.

Metatone. Thiamine-containing tonic.

Metenix (metolazone). Thiazide diuretic, used in hypertension and oedema.

Meterfolic (ferrous fumarate and folic acid). Iron and folic supplement to treat or prevent some forms of anaemia.

Metformin. Oral antidiabetic agent.

Methadone. Powerful analgesic.

Methenamine (hexamine). Urinary anti-infective agent.

Methicillin. Antibiotic of the penicillin group.

Methixene. Used in Parkinson's disease and related disorders.

Methocarbamol. For relieving muscle spasm after sprains.

Methohexitone. Anaesthetic injection.

Methotrexate. Anticancer drug.

Methotrimeprazine. Tranquillizer used in schizophrenia and also in severe physical illness.

Methoxsalen. Restricted drug for specialist use to increase the effect of ultraviolet radiation on the skin in psoriasis.

Methylcellulose. Bulk agent used to control bowel function in diarrhoea and constipation.

Methylcysteine. Expectorant used to reduce the viscosity of sputum.

Methyldopa. Antihypertensive.

Methylphenobarbitone (CD). Anticonvulsant.

Methylprednisolone. Strong steroid, used to reduce allergic reactions and other inflammatory diseases.

Methyprylone. Sedative, sleep inducer. Not usually recommended.

Methysergide. Prevents migraine attacks. For specialist use only.

Metipranolol. Beta-blocker eye drops used in glaucoma.

Metoclopramide. Antinauseant.

Metolazone. Thiazide diuretic, used in hypertension and oedema.

Metopirone. Blocks adrenal cortex secretion. Used in overfunction of adrenal gland.

Metoprolol. Beta-blocker. Many uses, including hypertension.

Metoros (metoprolol). For high blood pressure and angina.

Metosyn (fluocinonide). Very strong steroid for application to skin.

Metox (metoclopramide). Antinauseant.

Metramid (metoclopramide). Antinauseant.

Metriphonate. For bilharzia infections.

Metrodin (follicle-stimulating hormone). For infertility in females.

Metrolyl (metronidazole). Antimicrobial.

Metronidazole. Antimicrobial.

Metyrapone. Blocks steroid secretion by the adrenal glands. Used in Cushing's disease.

Mexiletine. Used for abnormal heart rhythms.

Mexitil (mexiletine). Used for abnormal heart rhythms.

Mianserin. Antidepressant.

Miconazole. For fungal infections.

Microgynon 30 (ethinyloestradiol and levonorgestrel). Combined oral contraceptive.

Micronor (norethisterone). Progesterone-only oral contraceptive.

Microval (levonorgestrel). Progesterone-only oral contraceptive.

Mictral (nalidixic acid). Antibacterial for urinary infections.

Midamor (amiloride hydrochloride). Diuretic which retains potassium in the body.

Midazolam. Injected benzodiazepine. For profound sedation during investigative or surgical procedures.

Midrid (isometheptene mucate, and paracetamol). For migraine attacks.

Migraleve (buclizine, paracetamol, and codeine—pink tablet; paracetamol and codeine—yellow tablet). For the pain and nausea of migraine attacks.

Migravess (effervescent tablets of metoclopramide and aspirin). For the pain and nausea of migraine attacks.

Migril (ergotamine, cyclizine, and caffeine). For the pain and nausea of migraine attacks.

Mildisom (hydrocortisone cream). For irritating skin diseases.

Minilyn (ethinyloestradiol and lynoestrenol). Combined oral contraceptive.

Minocin (minocycline). Tetracycline antibiotic.

Minocycline. Tetracycline antibiotic.

Minodiab (glipizide). Oral antidiabetic.

Minoxidil. For severe hypertension. Can cause hairiness.

Mintec (peppermint oil). For abdominal colic.

Mintezol (thiabendazole). For hookworm and roundworm infestations.

Miraxid (pivampicillin and pivmecillinam). Powerful antibiotic mixture.

Mithracin (plicamycin; mithramycin). Used in malignant disease to lower a raised blood calcium.

Mithramycin (plicamycin). Used in malignant disease to lower a raised blood calcium.

Mitobronitol. Used in some forms of leukaemia.

Mitomycin. Anticancer agent.

Mitoxana (ifosfamide). Anticancer agent.

Mitozantrone. Anticancer agent.

Mixtard 30/70. Mixture of prolonged-acting insulin used in diabetes.

Mobilan (indomethacin). Anti-inflammatory analgesic, used in rheumatic disease.

Mobyflex (tenoxicam). For rheumatism.

Modecate (fluphenazine decanoate). Long-lasting injection used in some forms of chronic mental illness, including schizophrenia.

Moditen (fluphenazine hydrochloride). Long-lasting injection used in some forms of chronic mental illness, including schizophrenia.

Modrasone (alclometasone dipropionate). Steroid cream and ointment.

Modrenal (trilostane). Inhibits adrenal cortex. Used in Cushing's disease.

Moducren (timol, amiloride, and hydrochlorothiazide). Mixture of beta-blocker and diuretics, for hypertension.

Moduret-25 (amiloride and hydrochlorothiazide). Diuretic mixture, used in hypertension and oedema.

Moduretic (amiloride and hydrochlorothiazide). Diuretic mixture, used in hypertension and oedema.

Mogadon (nitrazepam). Benzodiazepine. Induces sleep.

Molcer (docusate sodium). Ear drops to remove wax.

Molipaxin (trazodone hydrochloride). Antidepressant.

Monaspor (cefsulodin sodium). Antibiotic. Reserved for infections resistant to other drugs.

Monistat (miconazole). Antifungal, used for vaginal thrush.

Monit (isosorbide mononitrate). For angina pectoris.

Mono-Cedocard (isosorbide mononitrate). For angina pectoris.

Monotard (insulin zinc suspension). Long-acting insulin injection for diabetes.

Monotrim (trimethoprim). Antibacterial for urinary infections.

Monovent (terbutaline). Bronchodilator used in asthma and bronchitis.

Monphytol (undecenoic acid paint). For fungal infections of skin and nails.

Morhulin (cod-liver oil and zinc oxide ointment). For varicose ulcers, pressure sores, and minor wounds.

Morphine. Powerful analgesic of opioid group, used in severe pain.

Morsep (cetrimide, ergocalciferol, and vitamin A cream). For napkin rash.

Motilium (domperidone). Powerful antinauseant.

Motipress (fluphenazine and nortriptyline). Mixture of powerful tranquillizer and antidepressant. Not usually recommended.

Motival (fluphenazine and nortriptyline). Mixture of powerful tranquillizer and antidepressant. Not usually recommended.

Motrin (ibuprofen). Anti-inflammatory analgesic for rheumatic and other forms of pain.

Movelat. Cream containing heparin, adrenal extract, and salicylic acid, used to massage into the skin over a rheumatic site.

Moxalactam (latamoxef). Antibiotic.

MST Continus. Tablets of long-acting morphine for severe pain.

Mucaine (aluminium hydroxide, magnesium hydroxide, and oxethazaine). For dyspepsia.

Mucodyne (carbocisteine). Reduces sputum viscosity.

Mucogel (aluminium and magnesium hydroxides). Antacids for dyspepsia.

Multilind (nystatin in zinc oxide ointment). For candida infections of the skin.

Mupirocin. Antibacterial for application to skin.

Muripsin. Used to increase acidity in the stomach.

Mustine. Anticancer agent.

Myambutol (ethambutol hydrochloride). Antituberculosis drug.

Mycardol (pentaerythritol tetranitrate). For angina pectoris.

Mycifradin (neomycin sulphate). Antibiotic used to sterilize bowel before surgery.

Myciguent (neomycin sulphate ointment). For skin infections.

Mycota (undecenoate). Cream, powder, and spray for fungous infections of skin.

Mydriacyl (tropicamide). Eye drops to dilate pupil.

Mydrilate (cyclopentolate hydrochloride). Eye drops to dilate pupil.

Myelobromol (mitobronitol). Anticancer agent used in leukaemia.

Mygdalon (metoclopramide). Antinauseant and anti-emetic.

Myleran (busulphan). Anticancer agent used in leukaemia.

Mynah (ethambutol and isoniazid). Combined treatment for tuberculosis.

Myocrisin (sodium aurothiomalate). Gold injections for rheumatoid arthritis.

Myotonine (bethanechol). Used to stimulate bladder to empty.

Mysoline (primidone). Anti-epileptic.

Mysteclin (syrup contains tetracycline and amphotericin). Mixture of antibiotic and antifungal.

Nabilone. Derivative of cannabis for severe nausea and vomiting.

Nabumetone. Anti-inflammatory analgesic for rheumatic disorders.

Nacton (poldine methylsulphate). For intestinal spasm and peptic ulcer pain.

Nadolol. Beta-blocker. Many uses, including hypertension and prevention of migraine.

Naftidrofuryl oxalate. Used for blood vessel disease in brain and limbs.

Nalbuphine. Injected opioid analgesic. For severe pain.

Nalcrom (sodium cromoglycate). For colitis and for some forms of food allergy.

Nalidixic acid. Antibacterial for urinary infections.

Naloxone. Antidote for actions of opioids.

Nandrolone. Anabolic steroid to encourage protein build-up in debilitated patients.

Naprosyn (naproxen). Anti-inflammatory analgesic used in rheumatic disease.

Naproxen see *Naprosyn.*

Narcan (naloxone). Antidote for actions of opioids.

Nardil (phenelzine). Monoamine oxidase inhibitor anti-depressant.

Narphen (phenazocine hydrobromide). Opioid analgesic for severe pain.

Naseptin (chlorhexidine and neomycin). Nasal cream. For infections and for treating staphylococcal carriers.

Natamycin. For fungal infections.

Natrilix (indapamide). Weak diuretic with strong anti-hypertensive effect.

Natuderm. Emulsifying skin cream.

Natulan (procarbazine). Anticancer agent.

Navidrex (cyclopenthiazide). Diuretic used for hypertension and oedema.

Navidrex-K (cyclopenthiazide with potassium). Diuretic.

Naxogin (nimorazole). For acute ulcerative gingivitis.

Nebcin (tobramycin). Aminoglycoside antibiotic.

Nedocromil. Inhalation to prevent asthmatic attacks.

Nefopam. Analgesic for moderate pain.

Negram (nalidixic acid). Antibacterial for urinary infections.

Neocon-I/35 (norethisterone and ethinyloestradiol). Combined oral contraceptive.

Neo-Cortef (hydrocortisone acetate and neomycin sulphate). For infected dermatosis.

Neo-Cytamen (hydroxycobalamin). Vitamin B_{12} for pernicious anaemia and other similar deficiency states.

Neogest (norgestrel). Progesterone-only contraceptive.

Neo-Medrone (methylprednisolone). Steroid.

Neo-Mercazole (carbimazole). For thyrotoxicosis.

Neomycin. Aminoglycoside antibiotic.

Neo-NaClex (bendrofluazide). Diuretic used for hypertension and oedema.

Neosporin (gramicidin, neomycin, and polymyxin). Mixed antibiotic eye drops for bacterial eye infection.

Neostigmine. Increases the power of nerve stimulation to muscles. Used in myasthenia gravis.

Nepenthe. Morphine-containing solution. For severe pain.

Nephril (polythiazide). Thiazide diuretic, used for hypertension and oedema.

Nericur (benzoyl peroxide gel). For acne.

Nerisone (diflucortolone valerate). Strong steroid cream for severe inflammatory skin disorders.

Netillin (netilmicin). Reserved antibiotic for infections resistant to other drugs.

Netilmicin. Reserved antibiotic for infections resistant to other drugs.

Neulactil (pericyazine). Powerful tranquillizer for agitated forms of severe mental illness.

Neurodyne (co-codamol—paracetamol with codeine phosphate). Analgesic.

Nicardipine. Calcium antagonist for angina pectoris and hypertension.

Niclosamide. For human tapeworm.

Nicofuranose. Nicotinic acid derivative for peripheral vascular disease and to lower raised blood lipids.

Nicotinamide. Member of vitamin B group.

Nicotinic acid. For peripheral vascular disease and to lower raised blood lipids.

Nicotinyl alcohol. Nicotinic acid derivative. Vasodilator for peripheral vascular disease.

Nicoumalone. Anticoagulant.

Nidazol (metronidazole). Antimicrobial.

Nifedipine. Calcium antagonist for angina pectoris and hypertension.

Niferex (polysaccharide–iron complex). For iron-deficiency anaemia.

Nilstim (methylcellulose). Bulk agent for obesity.

Nimodepine. For certain types of stroke.

Nimorazole. Antimicrobial for trichomoniasis and ulcerative gingivitis.

Nimotop (nimodepine). For certain types of stroke.

Niridazole. For guinea-worm infection of the tissues.

Nitoman (tetrabenazine). Tranquillizer used in Huntington's chorea, tardive dyskinesia, and senile chorea.

Nitrazepam. Benzodiazepine, used to induce sleep.

Nitrocine (glyceryl trinitrate). Injection for angina pectoris and heart failure.

Nitrocontin Continus. Sustained-release glyceryl trinitrate tablets for angina pectoris.

Nitrofurantoin. Antibacterial for urinary tract infection.

Nitrolingual Spray (glyceryl trinitrate). Mouth spray for angina pectoris.

Nitronal (glyceryl trinitrate). Injection for angina pectoris and heart failure.

Nivaquine (chloroquine). For prophylaxis and treatment of malaria and for amoebiasis.

Nivemycin (neomycin). Tablets to sterilize bowel before intestinal operations.

Nizatidine. H_2 histamine antagonist, for peptic ulcers.

Nizoral (ketoconazole). Oral preparation for fungal infections.

Nobrium (medazepam). Benzodiazepine for the short-term suppression of anxiety.

Noctec (chloral hydrate). For the short-term treatment of insomnia.

Noltam (tamoxifen). Oestrogen antagonist, used in breast cancer.

Nolvadex (tamoxifen). Oestrogen antagonist, used in breast cancer.

Nonoxinol. Contraceptive spermicidal film used with a barrier contraceptive.

Noradran (diphenhydramine, diprophylline, ephedrine, and guaiphenesin). Cough preparation.

Noratex (cod liver oil and zinc oxide cream). Mild skin softener.

Nordisk Wellcome Infuser. Injection device to administer insulin to diabetic patients.

Nordox (doxycycline). Tetracycline antibiotic, used for prostatitis and for other infections.

Norethisterone. Female (progestogen) sex hormone. Used as contraceptive and in other forms of therapy.

Norflex (orphenadrine citrate). For Parkinson's disease and some other disorders of muscle tone.

Norgeston (levonorgestrel). Progesterone-only contraceptive.

Norgestrel (levonorgestrel). Progesterone-only contraceptive.

Noriday (norethisterone). Progesterone-only contraceptive.

Norimin ethinyloestradiol and norethisterone). Combined oral contraceptive.

Norinyl-l (mestranol and norethisterone). Combined oral contraceptive.

Noristerat (norethisterone enanthate). Injected progesterone contraceptive.

Normacol (sterculia). Bulk fibre laxative.

Normacol Antispasmodic (sterculia and alverine citrate). Used in irritable bowel syndrome.

Normacol Plus (sterculia and frangula). Bulk fibre laxative.

Normax (danthron and docusate sodium). Laxative.

Normetic (amiloride hydrochloride and hydrochlorothiazide). Diuretic mixture, for hypertension and oedema.

Normison (temazepam). Benzodiazepine used to induce sleep.

Nortriptyline. Antidepressant (tricyclic).

Norval (mianserin hydrochloride). Antidepressant.

Noscapine. Opioid, used to suppress dry or painful cough.

Novantrone (mitozantrone). Anticancer agent.

Novopen. Injection device, for administering insulin to diabetic patients.

Nozinan (methotrimeprazine). Tranquillizing drug used in schizophrenia and distressing physical illness.

Nubain (nalbuphine hydrochloride). Opioid used to relieve post-operative pain.

Nuelin (theophylline). Oral preparation for asthma and bronchitis.

Nu-K. Potassium chloride tablets.

Nulacin (mixed calcium and magnesium alkalis). Antacid for dyspepsia.

Numotac (isoetharine hydrochloride). Oral preparation for asthma and bronchitis.

Nurofen (ibuprofen). Anti-inflammatory analgesic.

Nu-Seals Aspirin (enteric coated aspirin). Slow-release aspirin with reduced chance of stomach irritation.

Nutraplus (urea cream). Hydrating and softening skin cream.

Nutrizym GR. Replacement for deficiency of intestinal secretions.

Nystadermal (triamcinolone acetonide and nystatin cream). For severe inflammatory skin disorders with candidal infection.

Nystaform (chlorhexidine hydrochloride and nystatin). For candidal infections.

Nystaform HC (hydrocortisone, chlorhexidine hydrochloride, and nystatin). For inflammatory skin disorders which are secondarily infected.

Nystan (nystatin). For candidiasis.

Nystatin. For candidiasis.

Nystavescent (nystatin effervescent pessaries). For vaginal thrush.

Octovit (mixed vitamins and minerals, including iron). No indication for this mixture, apart from dietary deficiency.

Ocusert Pilo (pilocarpine). Small tablet for application to the inside of the eyelid. For glaucoma.

Ocusol (sulphacetamide and zinc sulphate). Sometimes used for eye infections.

Oestradiol. Female sex hormone. Many therapeutic uses.

Oestriol. Female sex hormone. Many therapeutic uses.

Oestrogens see Oestradiol, Oestriol.

Oilatum. Cream and emollient to lubricate skin.

Olsalazine. For ulcerative colitis.

Omega-3 marine triglycerides. Fish oils used to lower raised blood lipids.

Omeprazole. For peptic ulcer.

Omnopon (papaveretum). Mixture of opium products, used for pain after operations.

Omnopon–Scopolamine. Mixture injected before surgical operations. Produces sedation, dries mouth and throat secretions, and reduces nausea.

Oncovin (vincristine sulphate). Anticancer agent.

One-alpha (alfacalcidol; 1α-hydroxycholecalciferol). Used to treat bone disease (osteomalacia) which may complicate renal failure.

Operidine (phenoperidine hydrochloride). Opioid used to relieve surgical pain.

Ophthaine (proxymetacaine hydrochloride). Local anaesthetic used on the surface of the eye.

Opiate squill linctus. Powerful suppressor of cough (and of breathing). Not usually recommended.

Opilon (thymoxamine). Used to improve blood flow in limbs and extremities.

Opioids. Group of drugs similar to morphine. Powerful analgesics but also are addictive and depress respiration.

Opium. Dried juice of seed chamber of oriental poppy. Contains many active substances, including morphine.

Opsite. Semipermeable, extensible film used as a post-operative dressing.

Opticrom (sodium cromoglycate). For allergic conjunctivitis.

Optimax (tryptophan). Used in depressive illness.

Optimine (azatadine maleate). Antihistamine, used in allergies.

Orabet (metformin hydrochloride). Oral antidiabetic drug.

Oradexon (dexamethasone). Strong anti-inflammatory steroid.

Oral-B. Fluoride. For preventing tooth decay.

Oralcer. Lozenge containing clioquinol. Antiseptic for mouth.

Oraldene (hexetidine). Mouthwash and gargle.

Orap (pimozide). Tranquillizer used in schizophrenia.

Orbenin (cloxacillin). Antibiotic of penicillin group, reserved for staphylococcal infections.

Orciprenaline. Bronchodilator used for asthma and bronchitis.

Orimeten (aminoglutethimide). For breast cancer.

Orphenadrine. For Parkinson's disease and other abnormalities of muscle action.

Ortho Dienoestrol (dienoestrol cream). For local applications to vagina if oestrogen is lacking.

Ortho-Creme. Spermicidal cream used with a barrier contraceptive.

Orthoforms. Spermicidal pessaries with a barrier contraceptive.

Ortho-Gynest (oestriol pessaries). For local application to vagina if oestrogen is lacking.

Ortho-Gynol. Spermicidal jelly used with barrier contraceptives.

Ortho-Novin 1/50 (mestranol and norethisterone). Combined oral contraceptive.

Orudis (ketoprofen). Non-steroidal anti-inflammatory analgesic used in rheumatic disease.

Oruvail (ketoprofen). Non-steroidal anti-inflammatory analgesic used in rheumatic disease.

Ossopan (hydroxyapatite). Calcium supplement.

Otosporin (hydrocortisone, neomycin sulphate, and polymyxin B sulphate drops). For infected and eczematous external ear canal.

Otrivine (xylometazoline hydrochloride). Nasal decongestant.

Otrivine-Antistin (xylometazoline and antazoline). Decongestant with antihistamine, used for nose and eye allergies.

Ovestin (oestriol). Oestrogen cream for application to vagina.

Ovran (ethinyloestradiol and levonorgestrel). Combined oral contraceptive.

Ovranette (ethinyloestradiol and levonorgestrel). Combined oral contraceptive.

Ovysmen (ethinyloestradiol and norethisterone). Combined oral contraceptive.

Oxamniquine. For schistosomiasis (bilharziasis).

Oxanid (oxazepam). Benzodiazepine for short-term relief of anxiety.

Oxatomide. Antihistamine, for allergies.

Oxazepam. Benzodiazepine for short-term relief of anxiety.

Oxerutins (rutosides). For chilblains.

Oxpentifylline. For peripheral vascular disease.

Oxprenolol. Beta-blocker. Many uses including hypertension and angina pectoris.

Oxybenzone. In some sunscreens.

Oxybuprocaine. Local anaesthetic used in eye.

Oxycodone suppositories. Opioid, for severe pain.

Oxymetazoline. Nasal congestant.

Oxymetholone. For stimulation of blood production.

Oxypertine. Tranquillizer used in schizophrenia.

Oxyphenbutazone eye ointment. Non-steroidal anti-inflammatory analgesic used for the local treatment of eye inflammation.

Oxytetracycline. Tetracycline antibiotic.

Oxytocin. Hormone used to contract the uterus and initiate labour.

Pabrinex. Injection of vitamins B and C.

Pacitron (tryptophan). Used in treatment of depressive illness.

Padimate O. In Spectraban, a sunscreen.

Palaprin Forte. Slow-release aspirin compound, for pain and fever.

Paldesic (paracetamol syrup). For pain and fever.

Palfium (dextromoramide). Opioid analgesic, for severe pain.

Paludrine (proguanil hydrochloride). For the prevention of malaria.

Pamergan P100 (pethidine and promethazine). Opioid and sedative antihistamine. For pain during labour.

Pameton (paracetamol and methionine). For pain and fever. The addition of methionine is an attempt to reduce the risk of an overdose of paracetamol.

Pamidronate disodium. For lowering blood calcium and for Paget's and some other bone diseases.

Panadeine (codeine phosphate and paracetamol (co-codamol)). Analgesic mixture for moderate pain.

Panadol (paracetamol). For pain and fever.

Pancrease (pancreatin). Digestive enzyme used when lacking in cystic fibrosis and other pancreatic diseases.

Pancreatin. Digestive enzyme used when lacking in cystic fibrosis and other pancreatic diseases.

Pancrex (pancreatin). Digestive enzyme used when lacking in cystic fibrosis and other pancreatic diseases.

Panmycin (tetracycline). Broad spectrum antibacterial.

Panoxyl (benzoyl peroxide). Gel and wash for acne.

Panthenol (pantothenic acid). Member of vitamin B group.

Pantothenic acid. Member of vitamin B group.

Papaveretum. Extract of opium, used for operative pain.

Paracetamol. For pain and fever.

Paracodol (effervescent co-codamol (codeine phosphate and para-cetamol)). For moderate pain.

Parahypon (paracetamol, codeine phosphate, and caffeine). For migraine pain.

Parake (codeine phosphate and paracetamol (co-codamol)). For moderate pain.

Paraldehyde. Used in severe epilepsy.

Paramax (paracetamol and metoclopramide). For pain with nausea.

Paramol (paracetamol and dihydrocodeine tartrate (co-dydramol)). For moderate pain.

Paraplatin (carboplatin). Anticancer agent.

Pardale (paracetamol, codeine phosphate, and caffeine). For migraine pain.

Parentrovite. Injection of vitamins B and C.

Parfenac (bufexamac). Cream for mild skin inflammation.

Parlodel (bromocriptine). For Parkinson's disease, but also has effects on reproductive system and on pituitary gland.

Parmid (metoclopramide). Antinauseant.

Parnate (tranylcypromine). Monoamine oxidase inhibitor—anti-depressant.

Paroven (rutoside). For chilblains.

Parstelin (tranylcypromine and trifluoperazine). Combination of mono-amine oxidase inhibitor and antipsychotic drug, for depression. Not usually recommended.

Pavacol-D (pholcodine). Cough suppressant.

Paxadon (pyridoxine). Member of vitamin B group. Used to prevent neuritis when receiving isoniazid. Also for some anaemias and deficiency states.

Paxalgesic (dextropopoxyphene and paracetamol). Mild painkiller.

Paxane (flurazepam). For short-term treatment of insomnia.

Paxofen (ibuprofen). Non-steroidal anti-inflammatory analgesic, used mainly in rheumatic disease.

Paynocil (aspirin and glycine). For pain and fever.

Pecram (aminophylline). For asthma.

Pemoline. Brain stimulant. Not recommended.

Penbritin (ampicillin). Antibiotic of the penicillin group.

Penbutolol. Beta-blocker, used with diuretic for hypertension.

Pendramine (penicillamine). For severe rheumatoid arthritis. Specialist use only.

Penicillamine. For severe rheumatoid arthritis. Specialist use only.

Penicillin G (benzylpenicillin). Injected antibiotic.

Penicillin V (pheonxymethylpenicillin). Oral antibiotic.

Penidural (benzathine penicillin). Oral antibiotic.

Penject. Injection device to administer insulin to diabetic patients.

Penmix (insulin cartridge). For diabetes mellitus.

Pentaerythritol tetranitrate. For angina pectoris.

Pentamidine. For pneumocystis pneumonia.

Pentasa (mesalazine). For ulcerative colitis.

Pentazocine. Opioid, for severe pain.

Pentostam (sodium stibogluconate). For leishmaniasis.

Pepcid PM (famotidine). H_2 antihistamine, for peptic ulcer.

Peppermint oil. For intestinal spasm.

Percutol (glyceryl trinitrate). Applied to skin for angina pectoris.

Perfan (enoximone). For heart failure.

Pergonal (follicle-stimulating hormone). For infertility.

Periactin (cyproheptadine hydrochloride). For allergies. Also stimulates appetite.

Pericyazine. Tranquillizer used in schizophrenia.

Permitabs (potassium permanganate). Solution for wounds and acute eczema.

Perphenazine. Tranquillizer used in surgery and in schizophrenia. Also antinausea and antihiccough actions.

Persantin (dipyridamole). Reduces blood coagulation.

Pertofran (desipramine hydrochloride). Tricyclic antidepressant.

Pethidine. Opioid analgesia, used for the pain of labour.

Petrolagar (emulsion of liquid paraffin). Purgative. Not usually recommended.

Pevaryl (econazole). Cream, lotion, powder for fungal infections.

Pharmalgen. Bee or wasp venom extract, for desensitization to stings.

Pharmorubicin (epirubicin hydrochloride). Anticancer agent.

Phenazocine. Opioid analgesic, for severe pain.

Phenelzine. Monoamine oxidase inhibitor antidepressant.

Phenergan (promethazine hydrochloride). Antihistamine for allergies.

Phenethicillin. Antibiotic of penicillin group.

Phenindamine. Antihistamine, for allergies.

Phenindione. Anticoagulant, used in venous thrombosis.

Pheniramine. Antihistamine, for allergies.

Phenobarbitone (CD). Barbiturate, used in epilepsy.

Phenol. Antiseptic. Dilute solutions used as a gargle.

Phenolphthalein. Purgative.

Phenoperidine. Opioid analgesic used in surgery.

Phenoxymethylpenicillin (penicillin V). An oral penicillin.

Phensedyl. Cough linctus containing codeine, ephedrine, and promethazine. Not usually recommended.

Phentermine. Amphetamine-like drug, used for short-term suppression of appetite.

Phenylbutazone. Anti-inflammatory analgesic. Reserved for specialist use.

Phenylephrine. In eye drops (which dilate pupil), in nose drops (which reduce congestion), and in 'cold cures'.

Phenylpropanolamine. Decongestant, used in 'cold cures'.

Phenytoin. Anti-epileptic.

Pholcodine. Opioid used to suppress cough.

Phosphate-Sandoz (sodium phosphate). To lower a raised blood calcium.

Phospholine Iodide (ecothiopate iodide). Eye drops for glaucoma. For specialist use only.

Phyllocontin Continus. Long-acting oral theophylline for asthma and bronchitis.

Physeptone (methadone hydrochloride). Opioid analgesic, for severe pain and for opioid dependence.

Physostigmine. Eye drops for glaucoma.

Phytex (salicylic acid paint). For fungal skin infections.

Phytocil (salicylic acid paint or undecenoate powder). For fungal skin infections.

Phytomenadione (vitamin K). Used to reverse the action of warfarin and some other anticoagulants.

Picolax (sodium picosulphate). Purgative.

Pilocarpine. Eye drops for glaucoma.

Pimafucin (natamycin). For candidiasis.

Pimozide. Tranquillizer for schizophrenia and some other serious mental disorders.

Pindolol. Beta-blocker, for hypertension, angina, and many other uses.

Pipenzolate bromide. For gastro-intestinal spasm.

Piperacillin. Antibiotic reserved for specific, resistant infections.

Piperazine. For threadworm and roundworm infestations.

Piperazine oestrone sulphate. Oestrogen replacement and other gynaecological treatment.

Piportil Depot (pipothiazine palmitate). Maintenance injection for schizophrenia.

Pipothiazine palmitate. Maintenance injection for schizophrenia.

Pipril (piperacillin). Antibiotic reserved for specific resistant infections.

Piptal (pipenzolate bromide). For gastro-intestinal spasm.

Piptalin (pipenzolate bromide). For gastro-intestinal spasm.

Pirbuterol. Tablets and aerosol for asthma and bronchitis.

Pirenzepine. Reduces stomach acid secretion. For peptic ulceration.

Piretanide. Diuretic used for hypertension.

Piriton (chlorpheniramine maleate). Antihistamine for allergies.

Piroxicam. Non-steroidal anti-inflammatory analgesic for rheumatic disorders.

Pitressin. Hormone used in some pituitary diseases. Also to reduce haemorrhage for oesophageal varices.

Pivampicillin. Antibiotic of penicillin group.

Pivmecillinam. Antibiotic used in urinary infections for salmonellosis.

Piz Buin (2-ethylhexyl *p*-methoxycinnamate and oxybenzone). Sun screen.

Pizotifen. Drug to prevent attacks of migraine.

Plaquenil (hydroxychloroquine sulphate). Prophylaxis and treatment of malaria.

Platet. Low dose aspirin to prevent heart attacks.

Plesmet (ferrous glycine sulphate syrup). For iron-deficiency anaemia.

Plicamycin (mithramycin). Used to lower raised blood calcium in advanced cancer.

Podophyllin. Paint used to treat warts and calluses.

Podophyllotoxin. For genital warts.

Poldine. For gastro-intestinal spasm and for peptic ulcers.

Pollinex. Allergen extract of grasses for hay fever desensitization.

Polybactrin (polymyxin). Antibiotic.

Polycrol. Antacid mixture containing aluminium hydroxide, magnesium carbonate and hydroxide, and dimethisone.

Polyestradiol. Oestrogen used to treat prostatic cancer.

Polyfax (polymyxin). Antibiotic ointment.

Polymyxin. Antibiotic.

Polynoxylin. For mild oral fungous infections.

Polysaccharide-iron complex. For iron-deficiency anaemia.

Polystyrene sulphonate resin. To lower a raised blood potassium.

Polytar (coal tar preparations). For psoriasis and chronic eczema.

Polythiazide. Diuretic used in oedema and hypertension.

Polytrim (polymyxin). Antibiotic eye drops.

Ponderax (fenfluramine hydrochloride). Appetite suppressant.

Pondocillin (pivampicillin). Antibiotic of the penicillin group.

Ponstan (mefenamic acid). Non-steroidal anti-inflammatory agent for rheumatic disorders.

Posalfilin (podophyllum resin). Local application for warts.

Potaba (potassium aminobenzoate). For diseases with fibrosis. Not of proven value.

Potassium canrenoate. Diuretic for specialist use in some forms of oedema.

Potassium chloride. Electrolyte supplement for potassium deficiency.

Potassium citrate. For relief of discomfort in urinary infections.

Potassium iodide. Used before operations for thyrotoxicosis. Reduces thyroid gland secretion.

Potassium permanganate. Solution for acute eczema.

Povidone-Iodine. For infections of skin and mucous membranes.

Practolol. Beta-blocker reserved for cardiac emergencies.

Pragmatar (ointment containing tar, sulphur, and salicylic acid). For acne and eczema.

Pralidoxime. Used in organophosphorus poisoning.

Praxilene (naftidrofuryl oxalate). For vascular disease.

Prazepam. Benzodiazepine for short-term relief of anxiety.

Praziquantel. For tapeworm infestation.

Prazosin. For hypertension. Also reduces bladder neck obstruction.

Precortisyl (prednisolone). Steroid for allergic and some other inflammatory disorders.

Predenema (prednisolone enema). For ulcerative colitis and other types of proctitis.

Predfoam (prednisolone aerosol foam). For ulcerative colitis and some other forms of proctitis.

Prednesol (prednisolone). Steroid for allergy and some other inflammatory diseases.

Prednisolone. Steroid for allergy and some other inflammatory diseases.

Prednisone. Steroid for allergy and some other inflammatory diseases.

Predsol (prednisolone eye and ear drops). For the local treatment of inflammation.

Preferide (budesonide). Strong steroid for skin diseases.

Prefil (sterculia). Fibre granules swallowed before a meal to reduce appetite.

Pregaday (ferrous fumarate and folic acid). For the treatment or prevention of anaemia due to deficiency of iron and folic acid.

Pregnavite Forte F (ferrous sulphate, folic acid, vitamin A, vitamin B complex). For the treatment or prevention of deficiency states.

Premarin (conjugated oestrogens). For oestrogen replacement and for other gynaecological conditions. A cream is also available for local oestrogen treatment.

Prempak C (conjugated oestrogens and norgestrel). For menopausal syndromes.

Prenylamine. For angina pectoris.

Prepidil. For inducing labour or abortions.

Prepulsid (cisapride). For acid reflux.

Prescal (isradipine). For high blood pressure.

Pressimune (antilymphocyte immunoglobulin). Immunosuppressant.

Prestim (timolol maleate and bendrofluazide). Beta-blocker and diuretic for hypertension.

Priadel (lithium carbonate slow-release tablets). To prevent recurrent bouts of depression and swings of mood.

Prilocaine. Local anaesthetic.

Primalan (mequitazine). Antihistamine for allergies.

Primaquine. For the treatment of benign tertian malaria.

Primidone. Anticonvulsant.

Primolut N (norethisterone). Progestogen for gynaecological disorders.

Primoteston Depot (testosterone ester). Long-acting androgen injection for hypogonadism in men.

Primperan (metoclopramide). Antinauseant.

Prioderm (malathion). For pediculosis and scabies.

Pripsen (piperazine). For threadworm and roundworm infections.

Pro-Actidil (triprolidine hydrochloride). Antihistamine for allergies.

Pro-Banthine (proantheline bromide). For gastrointestinal spasm. Also used for enuresis and to facilitate barium enemas.

Probenecid. For gout prophylaxis. Has also been used to increase the retention of penicillin in the body.

Probucol. To lower a raised blood cholesterol.

Procainamide. For some abnormal heart rhythms.

Procaine. Local anaesthetic.

Procaine penicillin. Sustained-action penicillin.

Procarbazine. Anticancer agent.

Prochlorperazine. Used in schizophrenia. Also used to treat vomiting and vertigo.

Proctofibe (bran tablets). Regulation of bowel function.

Proctofoam HC (hydrocortisone aerosol foam). For proctitis and other local inflammation.

Proctosedyl (ointment and suppositories containing hydrocortisone and framycetin). For anal inflammation.

Procyclidine. For Parkinson's disease.

Profasi (chorionic gonadotrophin). For female infertility.

Proflavine. Mild skin antiseptic.

Progesic (fenoprofen). Non-steroidal anti-inflammatory agent. Used in rheumatic disorders.

Progesterone. One of the female sex hormones. Many uses in gynaecology, including contraception.

Progestogens. Substances with progesterone-like activity. One of the constituents of oral contraceptives.

Proguanil. Malarial prophylaxis.

Progynova (oestradiol). For menopausal syndromes.

Prolintane. Brain stimulant. Not recommended.

Proluton Depot (hydroxyprogesterone). Injections which have been used for repeated abortion.

Promazine. Tranquillizer used for schizophrenia and other abnormal agitated mental disorders.

Promethazine. Antihistamine, used for allergies and motion sickness. Sedating, can induce sleep.

Prominal (methylphenobarbitone) (CD). Anticonvulsant.

Prondol (iprindole). Antidepressant.

Pronestyl (procainamide hydrochloride). For abnormal heart rhythms.

Propaderm (beclomethasone cream and ointment). Strong steroid, for severe inflammation of skin.

Propafenone. For abnormal heart rhythms.

Propain (codeine, paracetamol, diphenhydramine, caffeine). Analgesic, sedative, antihistamine and stimulant mixture. Not usually recommended.

Propamidine isethionate. Antibacterial eye drops.

Propantheline bromide. For gastro-intestinal spasm. Also used for enuresis and to facilitate barium enemas.

Propess (dinoprostone). For induction of labour or abortion.

Propine (dipivefrine hydrochloride). Eye drops for some chronic forms of glaucoma.

Propofol. Intravenous anaesthetic.

Propranolol. Beta-blocker. Used for hypertension, angina pectoris, and many other conditions.

Propylthiouracil. For thyrotoxicosis.

Prostigmin For myasthenia gravis.

Prostin E_2. To induce therapeutic abortion.

Prostin F_2 alpha. To induce therapeutic abortion.

Prothiaden (dothiepin hydrochloride). Tricyclic antidepressant. Mildly tranquillizing.

Prothionamide. Reserve drug for tuberculosis and leprosy.

Protriptyline. Tricyclic antidepressant.

Pro-Vent (theophylline capsules). For asthma and bronchitis.

Provera (medroxyprogesterone acetate). Progestogen used as a contraceptive. Also can act as an antitumour agent.

Pro-Viron (mesterolone). Androgen for hypogonadism in males.

Proxymetacaine. Local anaesthetic.

Prozac (fluoxetine). Antidepressive.

Pseudoephedrine. Used in asthma and excessive nasal secretion.

Psoradrate (dithranol cream). For psoriasis.

Psoralens. Increase the effect of ultraviolet radiation on the skin. Used in psoriasis.

Psoriderm (coal tar cream). For chronic eczema and psoriasis.

Psorin (dithranol and coal tar ointment). For psoriasis.

Pulmadil (rimiterol hydrobromide). For asthma and bronchitis.

Pulmicort (budesonide). Steroid aerosol inhalation for asthma and bronchitis.

Puri-Nethol (mercaptopurine). Anticancer agent used in acute leukaemia.

PUVA. Combined treatment with psoralens and long-wave ultraviolet radiation for psoriasis.

Pyopen (carbenicillin). Antibiotic reserved for specific types of infection.

Pyralvex (anthraquinone and salicylic acid oral paint). For painful mouth conditions.

Pyrantel. For roundworm, threadworm, and hookworm infections.

Pyrazinamide. Antituberculosis drug.

Pyridostigmine. For myasthenia gravis.

Pyridoxine (vitamin B_6). For deficiency states, isoniazid toxicity and some forms of anaemia. Has been used for depression and for nausea.

Pyrimethamine. Antimalarial.

Pyrogastrone (carbenoxolone sodium and antacid). For oesophageal inflammation.

Quellada (lindane). For pediculosis and scabies.

Questran (cholestyramine). For abnormally raised blood cholesterol.

Quinalbarbitone. Barbiturate. Induces sleep. Not usually recommended.

Quinapril. For hypertension.

Quinidine. For abnormal rhythms of the heart.

Quinine. For acute malaria. Also helps night cramps.

Quinocort (hydrocortisone and hydroxyquinoline cream). For skin inflammation.

Quinoderm with Hydrocortisone (also contains benzoyl peroxide). For acne.

4-Quinolones. A group of powerful antibacterials which includes ciprofloxacin and nalidixic acid.

Quinoped (benzoyl peroxide). For fungal skin infections.

Rabro (deglycyrrhizinized liquorice). For peptic ulcer.

Ramodar (etodolac). Non-steroidal anti-inflammatory agent, for rheumatoid arthritis.

Ranitidine. H_2 blocker which reduces gastric acid secretion. For peptic ulcer.

Rapifen (alfentanil). Short-acting opioid analgesic. For pain control in surgery.

Rapitard MC (biphasic insulin injection). Mixture of rapid-acting and prolonged-acting insulin for diabetes.

Rastinon (tolbutamide). Oral antidiabetic drug.

Rauwolfia alkaloids. Have been used in hypertension, but can cause depression.

Razoxane see *Razoxin*.

Razoxin (razoxane). Occasionally used in leukaemia.

RBC (antazoline, cetrimide, and calamine cream). Application for itching skin.

Redeptin (fluspirilene injection). Antipsychotic for schizophrenia.

Redoxon (ascorbic acid = vitamin C). For scurvy. May reduce incidence of respiratory infections.

Regulan (ispaghula husk). Bulk purgative, for regulating intestinal function.

Rehibin (cyclofenil). To stimulate ovulation in the treatment of female infertility.

Relaxit. Enema.

Relifex (nabumetone). Non-steroidal anti-inflammatory agent. For rheumatic diseases.

Remnos (nitrazepam). Benzodiazepine, for the short-term relief of insomnia.

Reproterol. For asthma and bronchitis.

Reserpine. Has been used for hypertension. Can cause depression.

Resonium A (polystyrene sulphonate resin). To lower a raised blood potassium.

Resorcinol. Has been used in acne preparations.

Restandol (testosterone tablets). For hypogonadism in males.

Retin-A (tretinoin). Local application for acne.

Retinol (vitamin A). For dietary deficiency.

Retrovir (zidovudine). For HIV infection in AIDS patients.

Revanil (lysuride). For Parkinson's disease.

Rheumacin LA (indomethacin). Non-steroidal anti-inflammatory analgesic, for rheumatic diseases.

Rheumox (azapropazone). Non-steroidal anti-inflammatory analgesic. For rheumatic diseases.

Rhinocort (budesonide). Steroid aerosol for hay fever.

Rhubarb mixture. Purgative.

Rhumalgan (diclofenac sodium). Non-steroidal anti-inflammatory analgesic for rheumatic diseases.

Riboflavine (vitamin B₂). For dietary deficiency.

Ridaura (auranofin). Gold compound for rheumatoid arthritis.

Rifadin (rifampicin). Antibiotic for tuberculosis and leprosy.

Rifampicin. Antibiotic for tuberculosis and leprosy.

Rifater (rifampicin and pyrazinamide). For tuberculosis.

Rifinah (rifampicin and isoniazid). For tuberculosis.

Rikospray (silicone or balsam). For urinary rash, colostomy care, and pressure sores.

Rimactane (rifampicin). Antibiotic for tuberculosis and leprosy.

Rimactazid (rifampicin and isoniazid). For tuberculosis.

Rimifon (isoniazid). For tuberculosis, given with other drugs.

Rimiterol. Aerosol for asthma and bronchitis.

Rinatec (ipratropium). For hay fever.

Ritodrine hydrochloride. Inhibits contraction of uterus during labour.

Rivotril (clonazepam). Benzodiazepine used in epilepsy.

Roaccutane (isotretinoin). Powerful oral drug for acne.

Ro-A-Vit (vitamin A). For dietary deficiency.

Robaxin (methocarbamol). For painful muscle spasm after injury.

Robaxisal Forte (methocarbamol and aspirin). For painful muscle spasm after injury.

Robinul (glycopyrronium bromide). For gastro-intestinal spasm.

RoC Total Sunblock (ethylhexyl *p*-methoxycinnamate). Sunscreen.

Rocaltrol (activated vitamin D). For osteomalacia.

Roferon-A (interferon alfa-2a). For Kaposi's sarcoma in AIDS patients and in hairy cell leukaemia.

Rohypnol (flunitrazepam). Benzodiazepine, for short-term treatment of insomnia.

Ronicol (nicotinyl alcohol). For peripheral vascular disease.

Roter (mixed antacids and bismuth). For dyspepsia. Not usually recommended.

Rutosides. Have been used in cramps, peripheral vascular disease. Not proved to be effective.

Rybarvin (adrenalin and atropine inhalant). For asthma and bronchitis.

Rynacrom (sodium cromoglycate). For hay fever.
Rythmodan (disopyramide). For abnormal heart rhythms.

Sabidal SR 270 (theophylline tablets). For asthma and bronchitis.
Salactol (salicylic acid paint). For warts.
Salazopyrin (sulphasalazine). For ulcerative colitis.
Salbulin (salbutamol tablets). For asthma and bronchitis.
Salbutamol. For asthma and bronchitis.
Salcatonin. For Paget's disease of bone.
Salicylic Acid. In local applications for fungal skin infections and to remove warts and calluses.
Saluric (chlorothiazide). Thiazide diuretic for hypertension and oedema.
Salzone (paracetamol syrup). For pain and fever.
Sandimmun (cyclosporin). Immunosuppressant, to prevent rejection of organ grafts.
Sandocal (effervescent calcium). For dietary deficiency.
Sando-K (effervescent potassium). For potassium depletion.
Sanomigran (pizotifen). Prevention of migraine attacks.
Saventrine (isoprenaline hydrochloride). Used to speed heart action.
Schering PC4 (levonorgestrel and ethinyloestradiol). Postcoital contraceptive.
Scheriproct (cinchocaine and prednisolone ointment and suppositories). For perianal inflammatory conditions.
Scopoderm (hyoscine adhesive patch). For motion sickness.
Scopolamine (hyoscine). Used before operations to produce drowsiness, dry secretions, and inhibit vomiting.
Secadrex (acebutolol and hydrochlorothiazide). Beta-blocker and diuretic for hypertension.
Seconal (quinalbarbitone sodium). Barbiturate. Has been used for insomnia but not usually recommended.
Sectral (acebutolol). Beta-blocker. Many uses, including hypertension and angina pectoris.
Securon (verapamil). Calcium antagonist. For angina pectoris, hypertension, and abnormal heart rhythms.
Securopen (azlocillin). Penicillin antibiotic. Reserved for specific infections.
Selegiline. For Parkinson's disease.
Selexid (pivmecillinam). Antibiotic for urinary infections and salmonellosis.
Selexidin (mecillinam). Injected antibiotic for serious infections.
Selora. Salt substitute, mainly potassium chloride.

Semi-Daonil (glibenclamide). Oral drug for diabetes.

Semitard MC (insulin zinc suspension, amorphous). Medium-action insulin injection for diabetes.

Semprex. Antihistamine for allergies such as hay fever.

Senna. Purgative.

Senokot (senna). Purgative.

Sential (hydrocortisone and urea cream). For inflammatory skin diseases.

Septrin (co-trimoxazole). Antibacterial combination.

Serc (betahistine hydrochloride). To reduce attacks of Ménière's giddiness.

Serenace (haloperidol). Tranquillizer used for schizophrenia and other excited mental illnesses.

Serophene (clomiphene citrate). For female infertility.

Serpasil (rauwolfia). Has been used for hypertension but can cause depression.

SH-420 (norethisterone acetate). Progestogen, used for some tumours.

Siloxyl (aluminium hydroxide and dimethicone). For dyspepsia.

Silver nitrate. For the short-term treatment of infected lesions.

Silver sulphadiazine. For infected burns.

Simeco (aluminium and magnesium antacid and dimethicone). For dyspepsia.

Simple eye ointment (yellow soft paraffin). To soften crusts and lubricate eyelids.

Simple Linctus. Soothing syrup for a dry cough.

Simple Ointment (wool fat, hard paraffin, and stearyl alcohol). Useful for chronic dry skin lesions.

Simplene (adrenalin eye drops). For some forms of chronic glaucoma.

Simvastatin. For lowering blood cholesterol.

Sinemet (levodopa with carbidopa). For Parkinson's disease.

Sinequan (doxepin). Antidepressant.

Sinthrome (nicoumalone). Anticoagulant.

Sintisone (prednisolone). Steroid, for allergies and some other inflammatory diseases.

Siopel (dimethicone and cetrimide cream). Barrier cream.

Slo-Indo (indomethacin, slow-release). Non-steroidal anti-inflammatory analgesic. For rheumatic disorders.

Slo-Phyllin (theophylline, slow-release)—for asthma and bronchitis.

Sloprolol (propranolol, slow-release). Beta-blocker. Many uses.

Slow-Fe (ferrous sulphate, slow-release). For iron deficiency.

Slow-Fe Folic (ferrous sulphate and folic acid). For dietary deficiency.

Slow-K (resin-bound potassium). For potassium deficiency.

Slow-Pren (oxyprenolol, slow-release). Beta-blocker. Many uses, including hypertension and angina pectoris.

Slow-Trasicor. Beta-blocker. Many uses, including hypertension and angina pectoris.

Sno Phenicol (chloramphenicol eye drops). For eye infections.

Sno Pilo (pilocarpine eye drops). For glaucoma.

Sno Tears (polyvinyl alcohol eye drops). For tear deficiency.

Soda mint tablets (sodium bicarbonate tablets). For dyspepsia.

Sodium acid phosphate. Phosphate supplement for some forms of osteomalacia.

Sodium amytal. Barbiturate. Induces sleep. Not usually recommended.

Sodium aurothiomalate. Gold injections for rheumatoid arthritis.

Sodium bicarbonate. Antacid, for dyspepsia and to alkalinize the urine.

Sodium cellulose phosphate. For raised blood calcium and to reduce calcium absorption from diet.

Sodium citrate. To alkalinize urine.

Sodium cromoglycate. For the prevention of some allergic disorders, in particular asthma.

Sodium fluoride. For prophylaxis of dental caries. Experimental use in osteoporosis.

Sodium fusidate. Antibacterial. Reserved for specific types of infection.

Sodium ironedetate. For iron-deficiency anaemia.

Sodium nitroprusside. Intravenous blood pressure-lowering infusion.

Sodium perborate. Mouthwash.

Sodium picosulphate. Purgative, used before bowel investigations.

Sodium salicylate. Non-steroidal anti-inflammatory analgesic. Used in rheumatic disorders.

Sodium stibogluconate. For leishmaniasis.

Sodium sulphate. Purgative.

Sodium tetradecyl sulphate. Irritant injection to sclerose varicose veins.

Sodium valproate. Anticonvulsant.

Sofradex (dexamethasone, framycetin, and gramicidin ear and eye ointment). For infected inflammation of the external ear canal and eye infections.

Soframycin ointment (gramicidin and framycetin sulphate). For infections of the eye and ear canal.

Soframycin tablets and injection (framycetin sulphate). Aminoglycoside antibiotic.

Solarcaine (benzocaine cream). Local anaesthetic cream for local pain.

Solis (diazepam). Benzodiazepine for the short-term treatment of anxiety and insomnia.

Soliwax (docusate in oil). Ear drops to soften wax.

Solpadeine (effervescent paracetamol and codeine). For pain.

Solprin (dispersible aspirin). Non-steroidal anti-inflammatory analgesic, for pain and fever.

Solu-Cortef (hydrocortisone injection). For severe allergies and some other inflammatory diseases and adrenal insufficiency.

Solu-Medrone (methylprednisolone). For severe allergies and some other immune or inflammatory diseases.

Solvazinc (zinc sulphate). For zinc deficiency.

Somatonorm 4IU (human growth hormone (somatrem)). For growth hormone deficiency.

Somatrem. For growth hormone deficiency.

Sominex (promethazine). Antihistamine marketed for occasional insomnia in adults.

Somnite (nitrazepam). Benzodiazepine, for the short-term treatment of insomnia.

Soneryl (butobarbitone). Barbiturate. Has been used for insomnia but not usually recommended.

Soni-Slo (isosorbide dinitrate). For angina pectoris.

Sorbichen (isosorbide dinitrate). For angina pectoris.

Sorbid SA (isosorbide dinitrate). For angina pectoris.

Sorbitrate (isosorbide dinitrate). For angina pectoris.

Sotacor (sotalol hydrochloride). Beta-blocker. Many uses, including hypertension and angina pectoris.

Sotalol. Beta-blocker. Many uses, including hypertension and angina pectoris.

Sotazide (sotalol hydrochloride and hydrochlorothiazide). Beta-blocker and diuretic mixture for hypertension.

Sparine (promazine hydrochloride). Tranquillizer used in severe agitation, particularly in the elderly.

Spasmonal (alverine citrate). Antispasmodic used for spastic colon (irritable bowel syndrome).

Spectinomycin. Antibiotic reserved for gonorrhoea.

Spectraban (contains padimate O). Sunscreen.

Spectralgen. Desensitizing vaccine containing extract of grass pollen for hay fever.

Spinhaler. Device used to produce an inhalable powder of sodium cromoglycate. For the prophylaxis of asthma.

Spiretic (spironolactone). Diuretic for specific types of oedema. Specialist use only.

Spiroctan (spironolactone). Diuretic for specific types of oedema. Specialist use only.

Spiroctan-M (potassium canrenoate). Diuretic for specific types of oedema. Specialist use only.

Spirolone (spironolactone). Diuretic for specific types of oedema. Specialist use only.

Spironolactone. Diuretic for specific types of oedema. Specialist use only.

Sporanox (itraconazole). For thrush and other fungal infections.

Sprilon (spray of dimethicone). For urinary rash, pressure sores, and irritation around colostomies and ileostomies.

Stabillin V-K (phenoxymethylpenicillin as potassium salt). Oral penicillin.

Stafoxil (flucloxacillin). Penicillin antibiotic reserved for specific infections.

Stanozolol. Anabolic steroid to build up tissues during debilitating illnesses. Also for bone marrow failure.

Staphlipen (flucloxacillin). Penicillin antibiotic reserved for specific infections.

Staycept (spermicidal jelly and pessaries). Used with barrier contraceptives.

STD (sodium tetradecyl sulphate). Irritant injection and sclerose varicose veins.

Stelazine (trifluoperazine). Tranquillizer used in schizophrenia and some other serious mental illnesses.

Stemetil (prochlorperazine maleate). Tranquillizer with powerful antinauseous actions.

Sterculia. Bulk agent, used to regulate bowel activity.

Ster-zac (triclosan). Skin disinfectant.

Stesolid (diazepam administered by injection or rectally). For anxiety during medical procedures and to treat epilepsy.

Stiedex (desoxymethasone). Moderately potent steroid for inflammatory conditions of the skin.

Stilboestrol. Synthetic oestrogen. Many uses in gynaecology, including menopausal syndromes.

Stomobar. Barrier cream used around colostomies and ileostomies.

Streptase (streptokinase). Injection to dissolve blood clots.

Streptokinase. Injection to dissolve blood clots.

Streptokinase–streptodornase. Local application to clean wounds.

Streptomycin. Aminoglycoside antibiotic.

Stromba (stanozolol). Anabolic steroid to build up tissues in debilitating illnesses. Also for bone marrow failure.

Stugeron (cinnarizine). For nausea and giddiness.

Sublimaze (fentanyl). Short-acting opioid used to control pain in surgical procedures.

Sulcralfate. For peptic ulcers.

Sudafed (pseudoephedrine hydrochloride). Decongestant for common cold, hay fever, and asthma.

Sudocrem (zinc oxide mixture cream). For napkin rash and pressure sores.

Sulconazole. For fungus infections of the skin.

Suleo-C (carbaryl). For lice.

Suleo-M (malathion). For lice or scabies.

Sulfametopyrazine. For urinary infections and bronchitis.

Sulfamylon (mafenide). For skin infections, particularly in burns.

Sulindac. Non-steroidal anti-inflammatory analgesic. For rheumatic disease.

Sulphacetamide. Eye drops for some forms of conjunctivitis.

Sulphadiazine. Antibacterial sulphonamide, used for meningococcal infections.

Sulphadimidine. Antibacterial sulphonamide used for urinary infections.

Sulphamethoxazole. Antibacterial sulphonamide. A component of co-trimoxazole.

Sulphamethazine (sulphadimidine). Sulphonamide for urinary infections.

Sulphasalazine. For ulcerative colitis and Crohn's disease.

Sulphaurea. For urinary infections.

Sulphinpyrazone. For prophylaxis of gout.

Sulphur. In local preparations for acne.

Sulpiride. Tranquillizer used in schizophrenia.

Sulpitil (sulpiride). Tranquillizer used in schizophrenia.

Sultrin (sulphonamide cream). For local gynaecological infections.

Suprefact (buserelin). Hormone inhibitor used in carcinoma.

Suramin. For filariasis.

Surbex T (vitamin B complex and vitamin C). For dietary deficiency.

Surem (nitrazepam). Benzodiazepine, for the short-term treatment of insomnia.

Surgam (tiaprofenic acid). Non-steroidal anti-inflammatory analgesic used for rheumatic diseases.

Surmontil (trimipramine). Tricyclic antidepressant.

Suscard Buccal (glyceryl trinitrate). For angina pectoris.

Sustac (glyceryl trinitrate, slow-release). For angina pectoris.

Sustamycin (tetracycline). Antibiotic.

Sustanon (testosterone). For replacement in androgen deficiency.

Symmetrel (amantadine hydrochloride). For Parkinson's disease. Also has some action in influenza.

Synacthen. Stimulates the adrenal glands to secrete cortisol. Used in inflammatory disease.

Synadrin (prenylamine). Prophylaxis of angina pectoris.

Synalar (fluocinolone acetonide). Potent steroid for inflammatory skin disorders.

Syndol. Analgesic mixture, including paracetamol and codeine.

Synflex (naproxen sodium). Non-steroidal anti-inflammatory analgesic, used in rheumatic disorders.

Synkavit (menadiol sodium phosphate). For vitamin K deficiency.

Synphase (ethinyloestradiol and norethisterone). Combined oral contraceptive.

Syntaris (flunisolide nasal spray). For hay fever.

Syntocinon (oxytocin). To induce labour and contract the uterus.

Syntometrine (oxytocin and ergometrine maleate). To contract the uterus after delivery in order to expel the placenta and stop bleeding.

Syntopressin (lypressin). For some types of pituitary deficiency.

Syraprim (trimethoprim). Antibacterial for urinary infections and bronchitis.

Sytron (sodium ironedetate). For iron-deficiency anaemia.

Tachyrol (dihydrotachysterol). Activated vitamin D for some forms of osteomalacia.

Tagamet (cimetidine). H_2-antihistamine for peptic ulcer.

Talampicillin. Antibiotic of penicillin group.

Talpen (talampicillin). Antibiotic of penicillin group.

Tambocor (flecainide acetate). For abnormalities of cardiac rhythm.

Tamofen (tamoxifen). Hormone antagonist used in breast cancer.

Tamoxifen. Hormone antagonist used in breast cancer.

Tampovagan Stilboestrol and Lactic Acid. Pessaries, for the local treatment of oestrogen lack.

Tancolin (dextromethorphan). Cough linctus for children. Not usually recommended.

Tanderil eye ointment (oxyphenbutazone). Local treatment of eye inflammation.

Tarcortin (coal tar extract and hydrocortisone). For chronic eczema and psoriasis.

Targocid (teicoplanin). Reserve antibiotic.

Tavegil (clemastine). Antihistamine for allergic disorders.

Tears Naturale (dextran and hypromellose eye drops). For tear deficiency.

Tedral (ephedrine hydrochloride and theophylline). For asthma and bronchitis.

Teejel (choline salicylate gel). For oral lesions.

Tegretol (carbamazepine). For epilepsy and trigeminal neuralgia.

Teicoplanin. Reserve antibiotic.

Temazepam. Benzodiazepine for the short-term treatment of insomnia and to relieve anxiety during medical and surgical procedures.

Temgesic (buprenorphine). Opioid analgesic for severe pain.

Tenavoid (meprobamate). For anxiety. Not usually recommended.

Tenif (atenolol). For high blood pressure and angina.

Tenoret 50 (atenolol and chlorthalidone). Beta-blocker and diuretic for hypertension.

Tenoretic (atenolol and chlorthalidone). Beta-blocker and diuretic for hypertension.

Tenormin (atenolol). Beta-blocker. Many uses, including hypertension and angina pectoris.

Tenoxicam. For rheumatism.

Tensilon (edrophonium chloride). Used for the diagnosis of myasthenia gravis.

Tenuate Dospan (diethylpropion hydrochloride) CD. Short-term aid to dieting for obesity.

Teoptic (carteolol hydrochloride). Beta-blocker eye drops for glaucoma.

Terazosin. For hypertension.

Terbutaline. For asthma and bronchitis.

Tercoda (codeine and terpin). Cough suppressant.

Terfenadine. Antihistamine for allergies.

Terodiline. Relaxes the bladder in the treatment of urinary frequency and incontinence.

Terolin (terolidine hydrochloride). Relaxes the bladder in the treatment of urinary frequency and incontinence.

Teronac (mazindol). Short-term aid to dieting for obesity.

Terpoin (codeine, terpin, and guaiphenesin). Cough suppressant.

Terra-Cortril (tetracycline and hydrocortisone). For infected inflammatory conditions of the skin.

Terramycin (oxytetracycline). Antibiotic.

Tertroxin (liothyronine sodium). For emergency treatment of thyroid underfunction.

Testosterone. Male sex hormone, used for hypogonadism.

Tetmosol (monosulfiram). For scabies, particularly in children.

Tetrabenazine. Used in movement disorders, including Huntington's chorea.

Tetrabid-Organon (tetracycline). Antibiotic.

Tetrachel (tetracycline). Antibiotic.

Tetracosactrin. Stimulates adrenal cortex to secrete cortisol.

Tetracycline. Antibiotic.

Tetralysal (lymecycline). Antibiotic.

Tetrex (tetracycline). Antibiotic.

T/Gel (coal tar shampoo). For psoriasis, eczema, and scaling of the scalp.

Thalamonal (fentanyl). Short-acting opioid analgesic used in surgery.

Theodrox (aminophylline and aluminium hydroxide). For asthma and bronchitis.

Theo-Dur (slow-release theophylline). For asthma and bronchitis.

Theophylline. For asthma and bronchitis.

Thephorin (phenindamine tartrate). Antihistamine for allergies.

Thiabendazole. For hookworm infestation.

Thiamine (vitamin B_1). For dietary deficiency.

Thiethylperazine. For severe nausea and vertigo.

Thioguanine. Anticancer agent for acute leukaemia.

Thiopentone sodium. Intravenous, rapidly acting anaesthetic.

Thioridazine. Tranquillizer for schizophrenia and other agitated forms of mental illness.

Thiotepa. Anticancer agent.

Thovaline (cod-liver oil and zinc oxide ointment). Emollient for skin.

Thymol Glycerin compound. Mouthwash for oral hygiene.

Thymoxamine. For peripheral vascular disease.

Thyroxine sodium. For thyroid underfunction.

Tiaprofenic acid. Non-steroidal anti-inflammatory analgesic for rheumatic disorders.

Ticar (ticarcillin). Antibiotic reserved for specific infection.

Ticarcillin. Antibiotic reserved for specific infection.

Tigason (etretinate). For severe psoriasis and some related skin diseases. Specialist use only.

Tilade (nedocromil sodium). For the prophylaxis of asthma.

Tildiem (diltiazem hydrochloride). Calcium antagonist for angina.

Timentin (ticarcillin). Antibiotic reserved for specific infections.

Timodine (dimethicone, hydrocortisone, nystatin, and benzalkonium chloride cream). For napkin rash.

Timolol. Beta-blocker. In the form of eye drops for glaucoma.

Timoped (tolnaftate cream). For skin infections, especially fungal.

Timoptol (timolol eye drops). For glaucoma.

Tinaderm-M (nystatin and tolnaftate cream). For fungous infections of the skin.

Tinidazole. Antibacterial, similar to metronidazole.

Tinset (oxatomide). Antihistamine, for allergies.

Tioconazole. For fungal nail infections.

Titralac (calcium carbonate). For high blood phosphate.

Tixylix (pholcodine and promethazine linctus). Sedative cough suppressant.

Tobralex (tobramycin eye drops). For eye infections.

Tobramycin. Antibiotic for local treatment of infections.

Tocainide. For abnormal heart rhythms.

Tocopherols. Vitamin E.

Tofranil (imipramine hydrochloride). Tricyclic antidepressant.

Tolanase (tolazamide). Oral antidiabetic drug.

Tolazamide. Oral antidiabetic drug.

Tolbutamide. Oral antidiabetic drug.

Tolectin (tolmetin). Non-steroidal anti-inflammatory analgesic, used for rheumatic disorders.

Tolerzide (sotalol hydrochloride and hydrochlorothiazide). Beta-blocker and diuretic combination for hypertension.

Tolmetin. Non-steroidal anti-inflammatory analgesic, used for rheumatic disorders.

Tolnaftate. For skin infections—particularly those due to fungi.

Tolu linctus. For cough.

Tonocard (tocainide). For abnormal heart rhythms.

Topal (alginic acid and antacids). For dyspepsia.

Topicycline (tetracycline solution). Local treatment for acne.

Topilar (fluclorolone acetonide). Strong steroid cream and ointment for severe inflammatory skin disorders.

Torecan (thiethylperazine). For severe nausea and giddiness.

Tosmilen (demacarium bromide eye drops). For glaucoma.

Trancopal (chlormezanone). For short-term treatment of anxiety and insomnia.

Trandate (labetalol hydrochloride). For hypertension.

Tranexamic acid. To stop some forms of bleeding.

Transiderm-Nitro (glyceryl trinitrate skin patch). For angina pectoris.

Transvasin (benzocaine, nicotinate, and salicylate cream). Local anti-rheumatic treatment.

Tranxene (clorazepate dipotassium). for the short-term treatment of anxiety.

Tranylcypromine. Monoamine oxidase inhibitor, for depression.

Trasicor (oxprenolol). Beta-blocker. Many uses, including hypertension and angina pectoris.

Trasidrex (oxprenol and cyclopenthiazide). Beta-blocker and diuretic combination for hypertension.

Travogyn (isoconazole). Cream and pessaries, for vaginal candida infections.

Trazodone. Antidepressant.

Tremonil (methixene hydrochloride). For Parkinson's disease.

Trental (oxpentifylline). For peripheral vascular disease.

Treosulfan. Anticancer agent.

Tretinoin. Local treatment for acne.

Tri-Adcortyl (triamcinolone, gramicidin, neomycin, and nystatin cream and ointment). Strong steroid with antimicrobials for severe inflammation of the skin.

Triamcinolone. Steroid for general or local suppression of inflammation.

Triamco (triamterine with hydrochlorothiazide). Diuretic.

Triamterine. Diuretic which causes potassium retention.

Triazolam. Benzodiazepine for the short-term treatment of insomnia.

Tribiotic (neomycin, bacitracin, and polymyxin aerosol spray). For skin infections.

Triclofos. For the short-term treatment of insomnia.

Tridesilon (desonide). Strong steroid for severe inflammatory skin disorders.

Trifluoperazine. Tranquillizer for schizophrenia and other excited mental illnesses. Also antinauseant and antivertigo.

Trifluperidol. Tranquillizer for schizophrenia and other excited mental illnesses. Also antinauseant and antivertigo.

Trilisate (choline magnesium trisalicylate). Non-steroidal anti-inflammatory analgesic for rheumatic disorders.

Trilostane. For Cushing's disease.

Triludan (terfenadine). Antihistamine for allergic disorders.

Trimeprazine. Antihistamine, used for allergies, to reduce itch and as a sedative.

Trimethoprim. Antibacterial, see *Trimogal, Trimopan*.

Trimogal (trimethoprim). Antibacterial used for urinary infections.

Trimopan (trimethoprim). Antibacterial used for urinary infections.

Trimovate (clobetasone, oxytetracycline, and nystatin cream and ointment). Moderately powerful steroid with antimicrobials for inflammatory skin disorders.

Trinordiol (levonorgestrel and ethinyloestradiol). Combined oral contraceptive.

TriNovum (norethisterone and ethinyloestradiol). Combined oral contraceptive.

Triogesic (paracetamol and phenylpropanolamine). Nasal decongestant.

Triominic (pheniramine and phenylpropanolamine). Nasal decongestant.

Triperidol (trifluperidol). Tranquillizer for schizophrenia and excited psychotic states.

Triplopen (benethamine penicillin). Long-acting antibiotic injection.

Tripotassium dicitratobismuthate. For peptic ulcer.

Triprolidine. Antihistamine, for allergies.

Triptafen (amitriptyline and perphenazine). Combination of tricyclic antidepressant and major tranquilliser. Not usually recommended.

Trisequens (oestriol, oestradiol, and norethisterone). Combined contraceptive with varying hormone dose during cycle.

Trobicin (spectinomycin). Antibiotic reserved for gonorrhoea.

Tropicamide. Eye drops to dilate pupil.

Trosyl (tioconazole). For fungal nail infections.

Tryptizol (amitriptyline). Tricyclic antidepressant.

Tuinal (amylobarbitone and quinalbarbitone). Barbiturate mixture for insomnia. Not usually recommended.

Two's Company (nonoxinol). Spermicidal pessaries to be used with condoms or other barrier.

Ubretid (distigmine bromide). To strengthen bladder emptying.

Ukidan (urokinase). To dissolve blood clots.

Ultrabase. Emulsifying cream for dry skin.

Ultradil (flucortolone cream and ointment). Moderately strong steroid for inflammation of skin.

Ultralanum (fluocortolone). Moderately strong steroid for inflammation of the skin.

Ultraproct (cinchocaine, clemizone, and fluocortolone ointment and suppositories). For pain and irritation in the anal region.

Ultratard, Human (insulin zinc suspension, crystalline). Long-acting insulin injection for diabetes.

Undecenoates. For fungal infections of skin.

Unguentum Merck (emulsifying cream). For dry skin and many other uses.

Uniflu Plus Gregovite C (caffeine, codeine, diphenhydramine, and vitamin C). For the common cold.

Unigest (aluminium hydroxide and dimethicone). For dyspepsia.

Unimycin (oxytetracycline). Antibiotic.

Uniphyllin Continus (slow-release theophylline). For asthma and bronchitis.

Uniroid (cinchocaine, neomycin, and polymyxin ointment and suppositories). For painful infected lesions in the anal region.

Unisomnia (nitrazepam). Benzodiazepine, for the short-term treatment of insomnia.

Univer (slow-release verapamil). Calcium antagonist, for hypertension.

Uriben (nalidixic acid). Antimicrobial for urinary infections.

Urisal (sodium citrate). To relieve the discomfort of urinary infections.

Urispas (flavoxate hydrochloride). For urinary frequency and incontinence.

Urofollitrophin (FSH). For female infertility.

Urokinase. To dissolve blood clots.

Uromide (sulphaurea). Sulphonamide for urinary infections.

Uromitexan (mesna). Used to reduce the toxic effects of some anticancer drugs on the bladder.

Ursodeoxycholic acid. For gallstones.

Ursofalk (ursodeoxycholic acid). For gallstones.

Uticillin (carfecillin sodium). For resistant urinary tract infections.

Utovlan (norethisterone). Progestogen. Many uses in gynaecology.

Valium (diazepam). Benzodiazepine, for short-term relief of insomnia and anxiety.

Vallergan (trimeprazine tartrate). Antihistamine for allergies and itching.

Valoid (cyclizine). For nausea and vertigo.

Valproate sodium. Anticonvulsant.

Vancocin (vancomycin). Antibiotic reserved for serious infections.

Vancomycin. Antibiotic reserved for serious infections.

Vansil (oxamniquine). For biharziasis.

Vascardin (isosorbide dinitrate). For angina pectoris.

Vascodon A (antazoline and naphazoline eye drops). For allergic conjunctivitis.

Vasogen (barrier cream). For napkin rash, pressure sores, and anal irritation.

V-Cil-K (phenoxymethylpenicillin). Oral penicillin.

Veganin (aspirin, paracetamol, and codeine). For pain.

Velbe (vinblastine sulphate). Anticancer agent.

Velosef (cephradine). Antibiotic.

Velosulin (neutral porcine insulin). Insulin injection for diabetes.

Ventide (salbutamol and beclomethasone inhaler). For asthma and bronchitis.

Ventolin (salbutamol). For asthma and bronchitis.

Vepesid (etoposide). Anticancer agent.

Veractil (methotrimeprazine). For schizophrenia, other mental illness, and for terminal care.

Veracur (formaldehyde gel). For warts.

Verapamil. Calcium antagonist for angina pectoris, hypertension, and abnormal heart rhythms.

Vermox (mebendazole). For threadworm, hookworm, and tapeworm infestations.

Verrugon (salicylic acid ointment). For warts and removal of hard skin.

Vertigon (prochlorperazine). For nausea and vertigo.

Verucasep (glutaraldehyde gel). For warts.

Vibramycin (doxycycline). Tetracycline antibiotic.

Vibrocil (dimethindene, neomycin, and phenylephrine). For nasal infections. Not usually recommended.

Vidarabine. Antiviral agent.

Vidopen (ampicillin). Penicillin antibiotic.

Vigranon B (vitamin B). For dietary deficiency.

Villescon (prolintane and vitamins B and C). So-called tonic. Not usually recommended.

Viloxazine. Antidepressant.

Vinblastine. Anticancer agent.

Vinca alkaloids. Anticancer agents.

Vincristine. Anticancer agent.

Vindesine. Anticancer agent.

Vioform-Hydrocortisone. Steroid and antimicrobial for skin inflammation.

Vira-A (vidarabine). Antiviral agent.

Virormone (testosterone). For male gonadal insufficiency.

Virudox (idoxurine). For herpes (cold sores).

Visclair (methylcysteine hydrochloride). To reduce sputum viscosity.

Viskaldix (pindolol and clopamide). Betablocker and diuretic for hypertension.

Visken (pindolol). Beta-blocker for hypertension and angina pectoris.

Vista-Methasone (betamethasone). Steroid for inflammation of nose, eye, and ear.

Vita-E (vitamin E). For deficiency.

Vitamins. For deficiencies.

Vitamin B_{12}. For deficiency, particularly failure to absorb vitamin B_{12} (as in pernicious anaemia).

Vivalan (viloxazine hydrochloride). Antidepressant.

Volital (pemoline). Stimulant. Not usually recommended.

Voltarol (diclofenac sodium). Non-steroidal anti-inflammatory analgesic, for rheumatic disorders.

Volumatic. Large-volume spacer device for facilitating the inhalation of salbutamol and beclomethasone.

Warfarin. Anticoagulant.

Welldorm (dichloralphenazone). For the short-term treatment of insomnia.

Wellferon (interferon alfa-N1). For some types of leukaemia.

White liniment (contains turpentine). Rubbed over areas of deep seated pain.

Whitfield's Ointment. For fungal infections of the skin.

Xanax (alprazolam). Benzodiazepine for the short-term treatment of anxiety.

Xipamide. Diuretic for hypertension and oedema.

X-Prep (sennosides). Purgative to prepare bowel for X-ray.

Xylocaine (lignocaine). Local anaesthetic.

Xylocard (lignocaine). Intravenous injection for abnormal heart rhythms.

Xylometazoline. Decongestant used in eye and nose drops.

Xyloproct (aluminium, zinc, and hydrocortisone ointment and suppositories). For inflammation and itch in the anal region.

Xylotox (lignocaine). Local anaesthetic injection.

Yomesan (niclosamide). For tapeworm infestations.

Yutopar (ritrodrine hydrochloride). Inhibits uterine contractions and arrests labour.

Z Span (zinc sulphate). For zinc deficiency.

Zaditen (ketotifen). For the prevention of asthma attacks.

Zadstat (metronidazole). Antimicrobial.

Zantac (ranitidine). H_2-antihistamine for peptic ulcer.

Zarontin (ethosuximide). Anti-epileptic for petit mal.

Zidovudine. For AIDS virus infection.

Zimovane (zopiclone). For short term treatment of insomnia.

Zinacef (cefuroxime). Antibiotic.

Zinamide (pyrazinamide). Antituberculosis drug.

Zinc oxide skin preparations. Astringent component of many ointments and creams for eczema, napkin rash, and pressure sores.

Zinc sulphate eye drops. For excessive tear production.

Zincomed (zinc sulphate). For zinc deficiency.

Zinnat (cefuroxine). Antibiotic.

Zirtek (cetirizine). Antihistamine for allergies such as hay fever.

Zocor (simvastatin). Lowers blood cholesterol.

Zoladex (goserilin). For prostatic cancer.

Zopiclone. For short term treatment of insomnia.

Zovirax (acyclovir). Antiviral drug for herpes simplex, herpes genitalis, and herpes zoster.

Zuclopenthixol. Tranquillizer for schizophrenia and related mental disorders.

Zyloric (allopurinol). For the prevention of attacks of gout.

Zymafluor (sodium fluoride). To increase the resistance of teeth to caries.

Index

254